Technology Transfer Out of Germany After 1945

Studies in The History of Science, Technology and Medicine
Edited by John Krige, European University Institute, Florence, Italy

Volume 1
Technological Change: Methods and Themes in the History of Technology
Robert Fox

Volume 2
Technology Transfer Out of Germany After 1945
Edited by Matthias Judt & Burghard Ciesla

Other Volumes in Preparation

Making Isotopes Matter: F.W. Aston and the Culture of Physics
Jeff Hughes

The Historiography of Science and Technology: Whose History? Whose Science?
Thomas Söderquist

Molecularising Biology and Medicine: New Practices and Alliances, 1930s–1970s
Soraya de Chadarevian & Harmke Kamminga

Entomology, Ecology and Agriculture: The Making of Scientific Careers in North America, 1885–1985
Paolo Palladino

Technology Transfer Out of Germany After 1945

Edited by
Matthias Judt
German Historical Institute, Washington DC, USA

and

Burghard Ciesla
Forschungsschwerpunkt Zeithistorische Studien, Potsdam, Germany

harwood academic publishers
Australia • China • France • Germany • India • Japan • Luxembourg
Malaysia • The Netherlands • Russia • Singapore • Switzerland • Thailand
United Kingdom • United States

Published in The Netherlands by Harwood Academic Publishers GmbH.

 Printed in Singapore.

Emmaplein 5
1075 AW Amsterdam
The Netherlands

British Library Cataloguing in Publication Data

Technology Transfer Out of Germany After
1945. — (Studies in the History of
Science, Technology & Medicine, ISSN
1024-8048; Vol. 2)
I. Judt, Matthias II. Ciesla, Burghard
III. Series
338.926094

ISBN 3-7186-5822-4

Contents

Series Introduction

Studies in the History of Science, Technology and Medicine is a new book series which aims to stimulate research in the field, concentrating on the twentieth century. It seeks to contribute to our understanding of science, technology and medicine as it is embedded in society, exploring the links between these subjects on the one hand and the cultural, economic, political and institutional contexts of their genesis and development on the other. Within this framework, and while not favouring any particular school of methodological approach, it welcomes studies which examine relations between science, technology and society in new ways, e.g. the social construction of technologies and large technical systems.

Contributors

Werner Abelshauser
Universität Bielefeld, Bielefeld, Germany

Mitchell G. Ash
University of Iowa, Iowa City, Iowa, USA

Johannes Bähr
Freie Universität Berlin, Berlin, Germany

Volker Berghahn
Brown University, Providence, Rhode Island, USA

Burghard Ciesla
Forschungsschwerpunkt Zeithistorische Studien, Postdam, Germany

Paul Erker
Freie Universität Berlin, Berlin, Germany

Jörg Fisch
Historisches Seminar der Universität Zurich, Zurich, Switzerland

Geoffrey J. Giles
University of Florida, Gainesville, Florida, USA

Matthias Judt
German Historical Institute, Washington, DC, USA

André Steiner
Universität Mannheim, Mannheim, Germany

Raymond G. Stokes
University of Glasgow, Glasgow, UK

Mark Walker
Union College, Schenectady, New York, USA

Preface

The collapse of communism in East Germany gave researchers access to archival sources whose very existence the GDR's authorities had denied for many years. The newly available sources enabled historians to examine long-standing claims, e.g., by the Soviets regarding reparations made by East Germany between 1945 and 1953, as well as previous western estimates of East German reparations. East Germany's economic collapse had raised questions about the underlying causes of its considerable economic lag. One explanation favoured by politicians and some historians referred to the different burdens East and West Germany had had to bear as regards dismantling and reparations.

In addition to the question of the material compensation paid by East Germany, new approaches influenced the historical debate on the reparations made by Germany after World War II. Among these new approaches, one of the most interesting was John Gimbel's attempt to describe Allied technology transfer from Germany and the migration of German scientists to Allied countries after 1945 as "intellectual reparations." This term, introduced by Gimbel in his pathbreaking book *Science, Technology, and Reparations, Exploitation and Plunder in Postwar Germany*,[1] argued that, by having scientists and intellectual property taken away by the Allies, Germany had made additional and hitherto unacknowledged reparations.

In September 1993, the German Historical Institute in Washington, DC, hosted a conference that dealt particularly with John Gimbel's approach. Entitled *Allied Technology Transfer from Germany or "Intellectual Reparations"?*, it asked whether the confiscation of documents in Germany and the hiring of German scientists by all the Allied powers could be considered as part of Germany's reparations or only as a special chapter in the already ongoing technology transfer between the countries involved. The conference was made possible by the generous support of the German *Fritz Thyssen Foundation* and gathered together scholars from Great Britain, France, Germany, Switzerland and the USA. Its proceedings have been supplemented in this volume by a

[1] Stanford CA: Stanford University Press 1990.

chapter written by three of the conference's commentators who have combined their remarks in an essay representative of the lively discussion at the meeting.

It emerged from the conference that the topic of Allied postwar policy towards Germany continues to be an interesting field of scientific discussion. The authors whose essays are published here approached this topic from different standpoints.

Volker Berghahn's chapter, with which we begin, questions, in principle, John Gimbel's attempt to regard Allied technology transfer from Germany as "intellectual reparations". After an appreciation of Gimbel's life-long work as a historian, Berghahn gives an introduction to the international discussion of technology transfer. Broadly speaking this debate assumes that countries that were technologically backward could benefit from the import of technology from more advanced nations. On this assumption, Germany would have been in a leading and the USA, for example, in a relatively backward position regarding technologies in 1945, if one followed Gimbel's theory.

Against this theory, Berghahn argues that by restricting his study to "a very special case of limited historical significance"[2] Gimbel overlooked the long history of German-American technology exchange. Just before the turn of the century the Germans' attitude towards the USA had changed dramatically, Berghahn explains. The USA, for a long time considered to be a land of "primitive unorganized capitalism", appeared at the end of the 19th century as a country in which speed, efficiency, courage and modernity were prime values, and were typically exemplified in standardized mass production.

Thus, Berghahn argues, it is no wonder that Germany began to look more carefully at the technological achievements of the USA. He mentions as examples the early translation of labour studies by Frederick W. Taylor into the German language and the extensive discussion in Germany of the effects of the assembly line production of cars. Even though the assembly line that first produced standardized cars initially raised significant opposition, it ultimately met with success.

Berghahn continues that West Germany became even more than ever dependent on American technologies after the war. In this connection, the more or less forced Allied technology transfer from Germany after 1945 looks rather like an exception to the rule. In contrast to Gimbel's general notion, Berghahn emphasizes rather that the USA had become the technological master of Germany rather than Germany that of the USA.

Several other authors add to Berghahn's analysis by investigating the effects of this technology transfer in Germany and the Allied countries. This is especially the case with the chapters by Raymond Stokes, Burghard Ciesla, Werner

[2] See: Volker Berghahn's paper "Technology, Reparations, and the Export of Industrial Culture. Problems of the German-American Relationship 1900–1950" published in this volume p. 4.

Abelshauser and André Steiner as well as with the comment made by Paul Erker, published here as part of the concluding chapter. Their essays focus on the effects of the technology transfer from Germany after 1945 on the development of industries and research in the German successor states and in the Allied countries.

Raymond Stokes aims to investigate the "normal case" of technology transfer in choosing Germany's chemical industry as the field for his investigation. He emphasizes that all former Allies undertook programs for the forced transfer of German chemical technology after 1945. Common among the Allies was the determination to include the removal of equipment and the confiscation of technical documentation. Also the capture or the hiring, respectively, of German scientists and technicians was part of these Allied programs. On the other hand, Stokes shows that the mix of these different means of transfer differed considerably between the western zones of occupation and the Soviet Zone. In this respect the case of the Soviet Zone in particular would need more detailed investigation.

Available studies and archival sources indicate that the impact of these Allied measures cannot be precisely assessed. However, Stokes points out that the measures undertaken by the Allies were neither as useful as the Allies hoped nor as devastating as the Germans feared. It is rather the case that the transfer of equipment, technical documentation and personnel to Allied countries helped to integrate the potential of the German successor states into the two newly established blocs, each of them dominated by one of the postwar superpowers. The chemical industries with their importance for both German postwar economies were significant for this process of integration.

The chapter by Ciesla explores the "special case" of German aeronautics, notably German windtunnel research. He suggests that at the end of the war German aeronautics were more advanced than their American counterpart in this field. During the years of the Nazi regime Germany had built most of the windtunnels the Americans found after 1945. The number and the quality of the windtunnels and of other air research facilities enabled German aircraft and rocket producers to have quick access to testing facilities. The contrary was true regarding American companies. Their research teams had "to wait in line" at American windtunnels. Ciesla points out that under these circumstances it was no wonder that American aircraft built by different companies looked similar.

The clear advantage the Germans had in the field of high-velocity testing facilities encouraged officials in the US Air Force to set up new American air research centers after the war. In the construction of these new facilities the Air Force used previous German experience. The copying also included the application of German experience in the organization of research.

Ciesla presents evidence to show that German scientists from the field of aeronautics were among the first and the largest groups of specialists brought to the USA by American occupation authorities. The transfer of German windtunnel technology, observed by the Americans to be more advanced, also demanded the hiring of German specialists.

Werner Abelshauser in his chapter investigates the long-term effects which forced technology transfer had for West Germany. He argues that the effects of Allied technology transfer were of limited importance. Even if employer associations claimed heavy losses from the removal of equipment and technical documents, their underlying purpose was to lobby for tax breaks and subsidies for the future reconstruction of factories and other facilities. According to Abelshauser's findings, these employer associations were not really concerned to seek compensation for reparation services, including the loss of intellectual property, by companies.

Although the confiscation of patents, blueprints and other documents meant a temporary disadvantage for West German companies, these firms could benefit from the immediate Allied policy. The confiscation of documents in Germany could serve as an indicator of the potential of West Germany's economy for other countries. This encouraged postwar cooperation based on contacts that had already existed in the interwar period.

There was no similar basis of existing, traditional structures of scientific cooperation that the migration of scientists and engineers to the Soviet Union could rely on, as André Steiner points out in his chapter. In contrast to scientists taken to the West, those taken to the USSR were not meant to stay in the victorious country for good. As a result East Germans in the Soviet Union had no reason for forging close links with the local scientific community. A kind of distrust of German scientists and engineers who had worked for the Soviet Union may also have been one reason for the relative lack of cooperation of science and research in COMECON.

Looked at from this vantage point, the "intellectual reparations" definitely had a punitive effect on East Germany, whereas the opposite became apparent for West Germany. East Germany could not develop an international community of scientists within the Eastern bloc based on the presence of German scientists in the Soviet Union. The West German case proves the opposite. The revitalization of scientific contacts between West Germany and the western Allies, also stimulated by the hiring of German scientists by Allied authorities after 1945, helped to reintegrate the West German economy into the world market.

Along with these papers, the analysis of Gimbel's theory at the Washington conference was also dealt with in talks that emphasized the judicial problems associated with the confiscation of documents in Germany after the end of World War II. Jörg Fisch analyzes this problem considering the international law on reparations that had emerged as a result of World War I. He discusses this not only as a problem of written international law but also in relation to the vital interests of the victorious powers.

Fisch gives a survey of the legal history of reparations. After World War I reparations replaced the pre-war concept of indemnities. According to the then emerging international law, reparations were defined as compensation for war damages. Fisch explains that even though the confiscation of intellectual property seems to appear as a kind of reparation, it is difficult to include these services in a real reparation account.

There are two reasons for this. First, the payment of reparations is based on an agreement between the parties involved indicating a certain amount of money and, connected to this, the definition of the kinds of reparation (percentage of trade, taxation, income or national revenue, kinds of machinery to be handed over). This is not the case with the transfer of intellectual property whose value is only in future use.

Second, in contrast with material goods, intellectual property is easy to hide or to destroy. The receiver of potential reparation deliveries in the form of intellectual property is not able to monitor whether he actually received the property demanded.

Fisch also comments on the special conditions after World War II that made the confiscation of intellectual property possible. In 1945 Germany had no government of its own that was able to negotiate the amount and the kind of reparation it had to make. Moreover, the unilateral power of the Allies enabled them to get access to German intellectual property. At the same time, the absence of international regulations regarding the seizure of documents and scientists excluded smaller victor countries from benefiting from this kind of "reparation".

Matthias Judt, in his essay, deals with the changing attitudes of the Allies towards occupied Germany. Apparently the Allied occupation powers had contradictory interests towards both Germany as the defeated nation and one another as victorious powers of World War II.

Judt argues that initially all the Allies were guided by a strong vision of Germany as the enemy in their wartime plans for the postwar treatment of the defeated country. This, and the fact that postwar investigations into the structure of Germany's economy confirmed wartime Allied conjectures, created an occupation policy that aimed for both the military and the industrial disarmament of Germany. Right from the beginning, this policy included demands for the dismantling or destruction of large portions of Germany's industrial facilities. In addition, the Allies also planned the capture of documents and what the Soviets called "specialists".

Although this was an expression of the Allies' hostile attitude towards Germany, at least the group of highly qualified scientists could benefit from the immediate Allied post-war regulations on research and development. Interested in the continuation of their professional careers, the specialists were looking for opportunities to work for Allied authorities. Judt presents evidence indicating that German scientists and technicians favoured employment by the Americans, but did not exclude a "Soviet option" regarding future employment. He goes on to show that, with the worsening of inter-Allied relations and their adoption of different political approaches, the policy of the Allies had already changed shortly after 1945. The exploitation of German potential in science and industry was then carried out by integrating them into newly established alliances rather than by dismantling them.

This policy change, which occurred in 1947–48, can be seen most obviously on the Soviet side. Soviet Corporations (SAG) were established instead of the planned dismantling of industrial facilities. The Soviet-owned Scientific Technical Bureaux in the eastern zone, first founded as collecting points for specialists who were to be transferred to the Soviet Union, continued to work fulfilling Soviet orders. Focusing on reparation deliveries out of current production, the Soviet Union created the demand eastern Germany needed to recover economically. A precondition for the willingness to place the focus on deliveries out of current production was the certainty that one would be able to keep troops in a part of Germany for a longer time than previously expected.

Judt adds that the western Allies also started to lift previously-implemented prohibitions and limitations on research and development in western Germany. The role of the former enemy nation Germany changed, and West Germany became an ally in the emerging Cold War. To improve West Germany's ability to finance necessary food imports and later to promote West German contributions to western European defence, production bans were lifted. Judt concludes that notwithstanding Germany's division into two states, the country as a whole was able to maintain most of its scientific and industrial potential even if the two German states were integrated into the eastern or western alliances, respectively.

In the pursuance of the Allied interests during the occupation of Germany, scientists tainted by their Nazi past could also benefit from uncertainties in Allied policy. Mitchell Ash and Mark Walker show in their chapters that the Allies had no clear concept about the denazification issue. Commenting, Geoffrey Giles confirms that the lack of a clear definition of what would count as denazification benefited the scientists who were tainted. Ash and Walker discuss this by choosing particular groups within German science. Ash concentrates his paper on genetics and psychology, whereas Walker deals with the case of German physicists.

Ash introduces his case studies with remarks on Allied policy regarding scientists who were of interest for the occupying powers. The Allies relied on "technocratic innocence",[3] as Ash calls the Allied approach, to overlook parts of the scientist's biographies. This notion assumes that, in the first place, science and technology was apolitical and value-neutral.

Ash explains that the Allies withdrew to this position for two reasons. First, there was their need to secure continuing university training in Germany, itself demanding enough teaching personnel that either had to remain in place or had to have a chance to be rehabilitated. Second, there was the wish to secure research and development in the interest of Allied authorities, this being done in Germany itself or even in the Allied countries.

[3] See Mitchell G. Ash: "Denazifying Scientists and Science", published in this volume, p. 61.

The case studies Ash presents explain how tainted scientists could present themselves as *Mitläufer* (collaborators) rather than as former active members of the Nazi party. The occupation authorities of the zone they were currently in accepted this picture painted by the former Nazis. Alternatively these scientists took advantage of the different approaches to the question in other zones. The eugenist Ottmar von Verschuer had already negotiated re-employment in Frankfurt when, with the help of a communist, Robert Havemann, American occupation authorities in Berlin put him under pressure. Finally von Verschuer got an appointment in Münster (British Zone). The psychologist Oswald Kroh tried to become employed in the Soviet Zone, an effort that failed, and finally was employed in the American Sector of Berlin. Another psychologist, Kurt Gottschaldt, successfully conducted his denazification in the Soviet Zone, but escaped from East Germany in 1962 after being put under pressure by the East German communist party, the SED. Ash's fourth case study deals with Hans Stubbe, describing how people that were not members of the NSDAP could serve as shields for others who had been party members.

Mark Walker's chapter analyzes the issue of denazification from the standpoint of German scientists. The chapter deals with the creation of the myth of "Nazi Physics" among German physicists after 1945. This myth contains the notion that only a few physicists had to be considered as Nazi physicists, whereas the majority of this professional group remained apolitical or even resistant towards the Nazi regime. However, the creation of a clear distinction within one professional group was used twice.

In the first phase, during the Nazi era, the majority described members of the so-called Deutsche Physik as radicals, dangerous for the maintenance of both the profession and the Third Reich. In denouncing the followers of Deutsche Physik, the other physicists could secure their own professional careers in showing loyalty to the Nazi regime. Walker calls the application of this method the renewed performance of the SA-Modell. This refers to the power struggle between the SS and the SA in 1934, when the SS successfully outmanoeuvred the SA during the Röhm affair.

In the second phase, the inter-professional struggles of the 1930s resurfaced when Allied authorities pushed for the denazification of the community of German physicists. Walker argues that the followers of Deutsche Physik served after 1945 as the scapegoats for many German physicists. By presenting the followers of Deutsche Physik as the "real Nazis" one had to fight with in the 1930s, the majority of German physicists, including those who were tainted with a Nazi past, could pretend that they had been resistant to the Nazis. In creating this impression, the majority of German physicists could prevent their own denazification.

The lack of understanding of the purpose of denazification on the Allied side is strongly supported by Geoffrey Giles who, in his jointly published comment on Ash's and Walker's chapters also argues that neither tries to define denazification themselves.

With the emerging Cold War concept of the new enemy, the denazification claim as such gradually became more qualified. This is obvious for the western Allies, but can also be noticed in the practice of the Soviet occupying power. It can be observed as a rule in the process of reintegrating former Nazis into the postwar societies of both East and West Germany, which continued to be determined by the Allies. Both German states were able to take advantage of both the Soviet Union's and the western Allies' changed policies.

Moreover German Nazi scientists even benefited from inter-Allied competition: in individual cases it was possible for one occupying power to lift "strict" denazification measures imposed by another. As for scientists taken to victorious Allied countries, however, the Allied authorities imposed a strict "isolation" for security reasons by placing them in remote areas of the respective countries.

All papers published in this volume question rather than support Gimbel's concept of "intellectual reparations". The main problem is to determine the "value" of technical documentation, patents, trade marks and above all, the German experts. It was possible to credit material deliveries to reparation accounts at current value, but the "value" of German intellectual property acquired by means of compulsory measures did not materialize before being applied in the victorious countries in a future time. That granted, Germany did suffer losses since the receiver nations did not compensate these confiscations by deliveries of goods in return — as is usual in other forms of technology transfer. Certainly, the receiver nations could save expenditure for similar developments in their countries by making use of previous German research results. Nevertheless one can endorse Christoph Buchheim's statement that any assessment and evaluation "made under economic aspects may provide some insight, but appears to be largely pointless from a political and legal point of view".[4]

---

[4] Buchheim, Die Wiedereingliederung Westdeutschlands in die Wellcristschaft 1945–1958 (München: R. Oldenbourg Verlag 1990) 77.

CHAPTER 1

Technology, Reparations, and the Export of Industrial Culture. Problems of the German-American Relationship, 1900–1950[1]

Volker R. Berghahn

The original impetus to hold this conference came from a book, entitled *Science, Technology, and Reparations, Exploitation and Plunder in Postwar Germany*, that the distinguished American historian John Gimbel published in 1990.[2] This book encouraged a number of younger scholars to explore in greater depth the questions that he had raised. They were also keen to add the Soviet dimension to the picture, now that the relevant archives have become available for the first time.[3] All this is meant to indicate that John Gimbel would have been at the center of this colloquium and his contributions to the discussion would have been invaluable. However, as we know, he is no longer with us. He died suddenly and all too prematurely on 16 July 1992 in his hometown of Arcata in California, a day after he had returned from Halle in Germany where, supported by a Fulbright grant, he had been researching his next book.

Until his death, John Gimbel, who was born in Hazelton, N.D. in 1922, had had a distinguished academic career, receiving his doctorate from the University of Oregon in 1956 and starting off as an instructor at Luther College, before moving on to take up appointments at such major institutions as the universities of Maryland and Indiana at Bloomington.[4] He also taught at two Canadian universities, Alberta and Saskatchewan, and finally settled for a professorship at Humboldt State University in California. He won many awards and distinctions, among them an honorary doctorate from Luther College and the Annual Book Prize of the AHA. He was named outstanding professor at Humboldt State and gained numerous research grants, among them from the ACLS, the NEH, the Rockefeller Foundation, and the Volkswagen-Stiftung. At age 70 and having retired from Humboldt State, he was still amazingly active and his fertile mind, I am sure, would have produced yet another important book. It was not to be. However, even if he was prevented from completing his life's work, he left us with what I would consider a very impressive oeuvre: a large number of articles and no less than four independent monographs, three of which were devoted to the history of Germany under Allied occupation after 1945. He ranks among the foremost experts on this period.

There is, to begin with, his *The American Occupation of Germany*, published in 1968,[5] which I regard as a classic and a pioneering study, an early example of the genre of grass-roots history that would sweep the historical profession a decade later. In this book, John Gimbel demonstrated how the all-too muddled policies of the United States affected the Germans. In looking at their plight, he displays a good deal of sympathy for the latter and hints that the training they were given in developing and running a democratic system was rather poor.

His *The Origins of the Marshall Plan*, published in 1976,[6] looked at the problem of reconstruction from a broader perspective and tried to explain how, with the unfolding of the Cold War, a liberal fear of fascism that had guided American policy in 1944–45 was replaced by a conservative fear of communism. Again, in the light of the proliferation of studies on the Marshall Plan during the 1980s, I think it is important to highlight the path-breaking role of this book.[7]

There is no doubt in my mind that John Gimbel has made an outstanding contribution to our understanding of what the United States intended to do, and did, in occupied Germany after World War II. He did so with a critical eye, possibly influenced by his first-hand experience as a young officer in postwar Germany. He questioned many of the myths about the politics and economics of the occupation period that had found their way into the autobiographies and memoirs of the major American actors. This is also true of his third monograph, *Science, Technology and Reparations*. Until he came along, the received wisdom had been that only the Soviets extracted heavy reparations from their zone of occupation. The Western Allies, by contrast, had learned the lesson from the post-1918 period and quickly abandoned the idea of forcing the Germans to make payments in cash and kind.[8] John Gimbel pointed to 'a much more comprehensive and systematic "intellectual reparations" program to exploit German scientific and technical know-how not only for military purposes, but also for the benefit of American science and industry.'[9] That he wanted these — as he called them — 'hidden' or 'creeping' reparations to be taken seriously as a historical problem is indicated by the blunt sub-title of his book: 'Exploitation and Plunder in Postwar Germany'.

To support his case, the author delved into the records of the US technical intelligence units which appeared in the factories and offices of German enterprises at the end of the war, searching for equipment, patents, lab test material and technical knowledge of potential usefulness to American industry. Although the conference will also deal with the 'Hunt for Nazi scientists', John Gimbel's study is more broadly based than Tom Bower's reproachful book on the so-called *Paperclip*

Conspiracy.[10] Now, I do not want to go into the details of *Science, Technology and Reparations*. I am sure it will be referred to many times in the course of this conference, while John Gimbel's oral contributions will be sorely missed.

I had the pleasure of meeting him only once at a conference some four years ago.[11] What impressed me most about him on that occasion was his openmindedness. As I indicated a moment ago, he never shied from going against the grain of received scholarship, and he clearly enjoyed the give-and-take of a good argument among historians. In the light of this, I am sure that he would have approved of my not stopping at this point, after I have celebrated his work and highlighted his achievements. Scholarship is an unending process. There are no final truths in historical writing; there is no definitive study of any topic of importance. We all make a contribution to an on-going debate. Some new findings will stand the test of criticism and time and hence join the increasingly complex pool of knowledge, providing a more sophisticated understanding of the problem at hand. Other arguments will be challenged and modified, if not even completely undermined by subsequent research.

It is in this spirit that I would now like to move a bit beyond *Science, Technology and Reparations* and raise some further questions that occurred to me as I re-read John Gimbel's work. I would like to suggest a framework within which the debates of this conference might fruitfully be seen. As I say this I see several worried faces, including that of Professor Abelshauser who, as you know, will be speaking about 'Der alliierte Technologietransfer — nur eine geringe Belastung Westdeutschlands bei seiner Eingliederung in the Weltwirtschaft'. So, let me assure all of you that I am not about to steal anybody's clothes. He and others, I assume, will be concerned with the *impact* of 'intellectual reparations' on the German and American economy. Rather I would like to do two things:

1. I would like to look at the pre-history to this conference's theme and examine the German-American technological relationship up to 1945.
2. I would like to advance the hypothesis that neither the pre-1945 relationship between the two countries nor what happened afterwards can be understood without conceiving of the role of technology in modern industrial societies in much broader terms than patents and machines.

Let me start off with this latter point and explain what I am thinking about.

Economic historians have, of course, long been interested in the links between economic growth and prosperity, or what some scholars have called 'economic progress', on the one hand and technology and technological creativity, on the other.[12] I am not reviving here the very old and basic debate concerning the difference between invention and innovation, the argument that the development of a machine by an ingenious individual will be futile unless it is seized upon by a dynamic innovator who applies the new technology and exploits it for commercial purposes.[13] Nor do I wish to engage with another controversy relating to the question of whether technological innovation is propelled by demand or induced by supply and capability.

Rather my focus is on the problem of technological diffusion within a particular national economy and, above all, across frontiers from one industrialized country to another. One aspect of this problem is whether in a capitalist economy there is such a thing as a 'free lunch'. Most economists, as you will know, have argued that there are no 'free lunches'. Writing on 'Economic Growth and Technological Progress' — my topic — Joel Mokyr has recently maintained that 'economic history is full of examples of free lunches as well as (more frequently) very cheap lunches', although, he admits, there are also 'endless instances of very expensive meals that ended up inedible and in some cases lethal'.[14] John Gimbel's *Science, Technology and Reparations* would appear to fall into the former category. His underlying assumption is that, through 'exploitation and plunder' (his subtitle) American industry got a free lunch on captured German technology. Without wishing to anticipate the results of this conference, many of you may well feel that John Gimbel was right. But on further deliberation, it began to dawn on me that he may have presented no more than a very special case of limited historical significance.

Now, the first point to be considered here is that the 'free lunch' argument has been applied most frequently to the notion of late-comers in the process of industrialization.[15] The assumption is that those 'backward' countries that were able to import technologies developed elsewhere reaped considerable advantages from their position of lesser development. They avoided the false starts and blind alleys of the technological leader; they saved on developments costs. At a time when Japan is often accused in the US of feasting without charge on American technologies and indeed is suspected of industrial espionage and 'stealing', it is well to remember that Jim Slater started his textile mill in Pawtucket, the cradle of the Industrial Revolution in America, with technologies that he had 'stolen' from Manchester in England.

Implicit in the notion of late-coming and technological diffusion is therefore the notion that there is a leading country whose superior position within the international economy triggers technological exports, legal or illegal, to more 'backward' nations, with the latter getting, if nothing else, a very cheap lunch. In the 19th century, Britain was the pioneer and undisputed hegemon in the unfolding saga of technological creativity and economic progress.[16] If it is plausible that, when we look at the industrialization process, the diffusion and flow of technology is largely one-way and, at this time, can be linked to the larger question of British hegenomy in the world, let me now add another factor, which emerges from the work of Heinz Hartmann, a Princeton-trained sociologist at Münster University in Germany.[17] Writing in the 1960s, he tried to expand the economic argument about technological exports by a cultural one. In his view it was impossible to study in isolation what he called instrumental exports from one country to another. Technology, he asserted, is always developed and applied in a particular socio-cultural context. An understanding of attitudes, mentalities, ways of organizing industrial work and generally of managing things is equally important. To Hartmann it is misleading to assume that technology is transferred to another country 'in the raw'. What is also transmitted is doctrines of production and distribution that were created in connection with it.

Before I try to apply this cultural notion of technological diffusion to the German-American relationship in the first half of this century, the other side, the importer of technology has to be considered as well, since the traditions and mindsets of the receiving industrial culture in turn act as filters. These filters will test how far the proposed imports are compatible with indigenous traditions. The compatibility question is likely to divide the business community. Some entrepreneurs will be more open to imports than others because they believe the underlying value systems to be more in tune with the national industrial culture than others. Although it is in principle conceivable that technologies may be imported with their underlying assumptions deleted, this — as far as I can see — does not apply to the German-American relationship. Its evolution was rather determined by power struggles within German industry which were eventually won by those who rejected American imports altogether and aimed to develop a special German industrial path.

I have now reached a point where I hope I have set up something like an analytical framework within which I would like to examine the German-American technological relationship up to 1945, thereby providing a historical backdrop to an analysis both of John Gimbel's pioneering work and of our proceedings at this meeting.

Up to the end of the 19th century, German industry which had seen a phenomenal growth in the previous three decades, perceived the United States generally like the rest of the German population: as a 'land of unlimited possibilities' perhaps, but also as the country of the 'Wild West', of a primitive unorganized capitalism, and of 'Red Indians' and 'racial chaos'.[18] But by the late 1890s, the tone of travel reports and tourist guides begins to change. America now appears as a country of speed, efficiency, courage, and modernity; a land where people are upwardly mobile, but also work very hard, frequently lose their job, enjoy no social security, and are exposed to racial discrimination, urban crime, and political corruption. German entrepreneurial reactions to the American pavillion at the Paris World Exhibition of 1900 present a good case in point. In particular the demonstration of new steel-making techniques mesmerized the German visitors, also — and this is important for my purposes — because they seemed to open up possibilities for standardized mass production. Soon scores of engineers and other experts could be seen travelling across the Atlantic to inspect the centers of advanced industrial production along the East Coast, in Ohio, Michigan, and Pennsylvania.

Sooner rather than later, they would all come across the name of Frederick W. Taylor and his movement.[19] Taylor's 1906 essay 'On the Art of Cutting Metals' was quickly translated. His work on 'Shop Management' appeared in Germany in 1909. It took only a year after the publication in 1911 of Taylor's most famous piece, entitled 'Principles of Scientific Management', that the book was translated into German. In the spring of 1913 it had to be reprinted three times. Now, unfortunately, there is no time for me to go into details of the pre-1914 German-American industrial relationship. But three points emerge in my view quite clearly:

1. The relationship was not merely about technologies, about metal-cutting techniques; it was also about work organization and all those 'principles of Scientific Management' that all experts realized were inseparable from the technologies developed on the other side of the Atlantic.
2. The relationship became increasingly hegemonic. With the exception of the German chemicals industry, which became a world leader especially in the development of dye-stuffs, the major branches were looking towards America.[20] Scores of employees were sent to America by their companies. As far as I can make out, there were no similar streams of industrial visitors coming to the Ruhr, to Stuttgart, or to Berlin.
3. The relationship encountered opposition, not just on the part of intellectuals and politicians who perpetuated the 19th-century stereotypes about the 'Wild West' and a primitive immigrant society without *Kultur*. Resistance also came, albeit from a very different perspective, from entrepreneurs who recognized that the importation of American technology carried with it the pressure to accept certain ways of organizing and running their industry.

An example of how all three of the aspects just mentioned were interconnected is provided by the case of the Daimler-Benz company, makers of the famous Mercedes cars. Next to Taylorism Ford's experiments with assembly-line production and the manufacture of a cheap standardized car, his T-model, had inevitably attracted a good deal of attention also in Europe. But the automakers of Stuttgart had their doubts, not for technical reasons, but for socio-cultural ones. As Daimler-Benz put it somewhat defiantly:[21] 'Over here we are still a long way from the American situation where every Mr. Jones owns a car. With us the automobile is for the most part a vehicle for the better-off classes.' So they continued to produce their luxury cars, announcing proudly: 'Here [we do things] meticulously and thoroughly; over there [in America it is] skimping and rushing.' Clearly Daimler-Benz was wary of the democratizing consequences of Americanization which *Der Motorwagen* had announced as early as 1905 was underway in Germany 'at an accelerated pace'.[22]

The experience of the electrical engineering firm of Robert Bosch just down the road from Daimler-Benz presents another facet of the hegemonic pull that industrial America exerted toward Germany as early as prior to World War I. Bosch had been keenly interested not only in American technology, but also in Taylor's ideas on factory organization and modern management. In 1913, H. Borst, a member of the Bosch board of directors, even succeeded in having a long interview with Taylor. When Borst wanted to expand the experiments in modern production that they had engaged in since the middle of the decade, he encountered the vigorous opposition of the local Metal Workers' Union to which many Bosch employees belonged.[23] The latter had become increasingly resentful of what had become known as the 'Bosch tempo' of rationalized production, and suddenly in 1913 Bosch had a major strike on their hands. What, for my purposes, is no less interesting is that many of Bosch's colleagues and competitors who had been uneasy about mass production techniques gloated when the Stuttgart firm suffered from the misfortune of this strike. They felt confirmed in their view that the importation of American ideas of production technology and organization brought nothing but trouble.

World War I disrupted most of the links that had been forged between German and American industry until 1914. But the war and the subsequent period of civil war, economic chaos and inflation was barely over, when German industry began to look toward the United States once more. From the start there were great hopes that America's superior financial power might help the country to reconstruct its war-ravaged industries. And indeed, after 1923 and with the American-brokered Dawes Plan, which settled the thorny reparations issue, in place, funds began to flow into Germany.[24] By 1928, capital exports had reached some $1.4 billion. Another $1.6 billion were given in short-term loans. Close to 80 American firms established branches in Germany or took a stake in indigenous enterprises. Ford built an assembly plant at Cologne. General Motors bought up Opel Cars, the first German maker of a cheap mass produced car, the *Laubfrosch*. Chrysler established a factory near Berlin. IG Farben and Zeiss Opticals signed cooperative agreements with American firms.[25]

That the United States had more than ever before become the embodiment of industrial modernity is evidenced by the fact that German entrepreneurs and engineers travelled across the Atlantic in larger and larger numbers. As P. Riebensahm, an engineering professor at Berlin, reported:[26] 'At first a few leading personalities came individually; then major firms sent their employees in groups of twos and threes.... Soon the passenger lists of the beautiful ships of Hamburg-America Line and of North German Lloyd looked like a register of the leading industrial forms of Germany.' Gustav Krupp, one of the visitors, wrote to a colleague:[27] 'I gained valuable impressions in America.... I can only advise you also to see that country once again.' IG Farben's Carl Bosch sent Wichard von Moellendorff on an extended tour in 1928 'to examine the transferability of experiences to Germany.'[28] Meanwhile the *Gutehoffnungshütte* steel trust, following a visit by its director general Paul Reusch, dispatched one of its engineers who wrote a 250-page report on his impressions of Ford and other factories, appending copious drawings of machines, but significantly also of plant layouts, diagrams and other materials.[29] No doubt, there was much to be discovered at Ford's River Rouge plant with its 75,000 workers and 27 miles of conveyor belts. So, I ask you to think of the meaning of all this: does it not demonstrate my point that the German-American industrial relationship was primarily one way and that technological exports were inseparable from exports of an immaterial kind that, just as before 1914, aroused enthusiasm and opposition.

The Great Slump and the rise of National Socialism in the 1930s put a heavy damper on German-American relations in general. But it did not disrupt industrial cooperation. Others at this conference are in a much better position than I am to talk about the links between the German and American chemical industry in the 1930s. Once the worst of the depression was over, visits to the centers of American technology and production resumed. In 1937 North German Lloyd advertised over 40 tours to see industrial America. Ferdinand Porsche, the father of the *Volkswagen*, had been over the year before to study Ford's plants at Michigan.[30] And as late as the spring of 1939 the German Labor Front organized no less than three trips to study the US economy. In short, interest in American technology and industry continued to be intense, but the ideological and political climate for the reception of American exports, whether instrumental or cultural, had, of course, worsened considerably. The ultimate ambition of Nazi Germany was to go it alone, not just power-politically and militarily by establishing a vast continental empire, but also technologically-industrially. In the future it was no longer a question of studying and learning from industrial America, but rather of outpacing it by producing bigger and better goods than the rival across the Atlantic.[31] Nazi competition with American technological feats is particularly intriguing with regard to architecture: a railroad station is Berlin bigger than Grand Central in New York; a bridge spanning the Elbe River near Hamburg bigger than Golden Gate Bridge in San Francisco — these were the visions of the Third Reich and its '*Führer*'.[32] He never got much beyond the planning stage before losing World War II from which the United States emerged as the clear technological leader among the industrial nations of the world.

And this takes me back to my introduction and my appreciation of John Gimbel's *Science, Technology and Reparations*. Again, without wishing to anticipate any results that may emerge from this conference, if one considers the German-American relationship up to 1945, it looks rather more that the United States became Germany's technological and industrial master. Nor did this relationship see any reversal after World War II.[33] On the contrary, postwar Germany became more than ever dependent not only on American financial support and investments, but also on the supply with machinery. Fordism came for a second time, and it arrived on the scene both in the shape of technological imports and of cultural practices which German entrepreneurs adopted. Once again German managers travelled to the United States in large numbers. And those who could not go were put into the Training-Within-Industry program which was sponsored by none other than Paul Hoffman, a former president of the Studebaker Corporation and now Administrator of the Marshall Plan.

John Gimbel wrote an important book and, as I said at the beginning, is sorely missed at this meeting. But in the light of the history of the German-American industrial relationship, I wonder whether it might not have been better for him to take his subtitle from a statement of his book. At one point he refers to the FIAT program as a 'conveyor-belt for the future business connections''[34] I think he was absolutely right. It fills me with great sadness that I can no longer ask him whether he agrees.

NOTES

1 Originally presented as a lecture whose style has not been changed.
2 Stanford, 1990.
3 See the article by Burghard Ciesla, pp. 93–106.
4 The following is based on information provided by the chairman of the Department of History, Humboldt State University. I would like to thank Professor Tanner for his help in this matter.
5 Stanford, 1968.
6 Stanford, 1976.
7 See, e.g., A. Braehler, *Der Marshallplan* (Cologne, 1983); S. Hoffman and C.S. Maier, eds., *The Marshall Plan* (Boulder, 1984); M.J. Hogan, *The Marshall Plan* (Cambridge, 1987); G. Bischof and C.S. Maier, eds., *The Marshall Plan and Germany* (Oxford, 1991); H.-J. Schroeder, ed., *Marshallplan und westdeutscher Wiederaufstieg* (Stuttgart, 1990).
8 See, e.g., N. Balabkins, *Germany Under Direct Controls* (New Brunswick, 1964); J. Foschepoth, ed., *Kalter Krieg und Deutsche Frage* (Göttingen, 1985). O. Nuebel, Die amerikanische Reparationspolitik gegenüber Deutschland (Frankfurt, 1980).
9 J. Gimbel, *Science*, vii.
10 Boston, 1987.
11 The conference, held at Harvard University, was concerned with the American impact on Japan, Austria, and Germany after 1945.
12 See, e.g., W.W. Rostow, The Stages of Economic Growth (Cambridge, 1960); D. Landes, *Unbound Prometheus* (Cambridge, 1969).
13 See, e.g., J. Schumpeter, *The Theory of Capitalist Development* (Cambridge, Mass., 1934).
14 J. Mokyr, *The Lever of Riches, Economic Growth and Technological Progress* (Oxford, 1990), 3f.
15 See, e.g., A. Gerschenkron, *Economic Backwardness in Historical Perspective* (Cambridge, Mass., 1962).
16 See, e.g., P. Mathias and J.A. Davies, eds., *Innovation and Technology in Europe* (Oxford, 1991).
17 H. Hartmann, *Amerikanische Firmen in Deutschland* (Cologne, 1963).
18 See, e.g., W. Helbich, "Different, but not out of this world: the German image of the United States between two wars, 1871–1914 (unpubl. MS, 1992).
19 See, e.g., H. Homburg, "Anfänge des Taylorsystems in Deutschland vor dem Ersten Weltkrieg", *Geschichte und Gesellschaft 2* (1978): 170–94; L. Burchardt, "Technischer Fortschritt und sozialer Wandel," in W. Treue, ed., *Deutsche Technikgeschichte* (Göttingen, 1977), 52–98.
20 See note 19 above.
21 Quoted in A. Kugler, "Von der Werkstatt zum Fliessband", *Geschichte und Gesellschaft* 13 (1987): 316.
22 Quoted *ibid.*, 313.

23 H. Homburg (note 19 above), 180ff.
24 See W. Link, *Die amerikanische Stabilisierungspolitik in Deutschland* (Düsseldorf, 1970)
25. See P. Hayes, *Industry and Ideology* (Cambridge, 1987).
26 Quoted in: M. Nolan, *Visions of Modernity* (Oxford, 1994), 18.
27 Ibid.
28 Quoted in: V.R. Berghahn, *The Americanisation of West German Industry, 1945–1973* (Oxford, 1986), 29.
29 See M. Nolan (note 26 above), 18f.
30 Ibid., 233.
31 See H.D. Schäfer, "Amerikanismus im Dritten Reich", in M. Prinz and R. Zitelmann, eds., *Nationalsozialismus und Modernisierung* (Darmstadt, 1991), 199–215.
32 See J. Thies, Architekt der Weltherrschaft (Düsseldorf, 1976).
33 See V.R. Berghahn (see note 28).
34 J. Gimbel (see note 2).

CHAPTER 2

Reparations and Intellectual Property

Jörg Fisch

After World War II, the victors systematically appropriated intellectual property in Germany which had scientific, technical, military, commercial or economic value.[1] Never before had this been done to a noteworthy degree following a war.[2] To what can this innovation be attributed?

The answer seems apparent. Never before had there been so much emphasis placed on technical knowledge and skills, research and development and the race toward more and more effective armaments as in World War II. This was manifested most conspicuously in the atomic bomb. If one wanted to ensure victory, one had to try to ensure himself superiority in the technical and scientific arena.

However, technical knowledge had significant influence long before 1945 on the military and — particularly — economic strength of nations. The objectification of this knowledge certainly gained greater and greater significance as a result of increasing amounts of technical literature and, more specifically, the introduction of the patent system in modern times. All this did not begin in 1945, however.

The fact that no intellectual property was appropriated in connection with peace settlements prior to 1945 is therefore probably due less to the lack of material to appropriate than to the barriers which existed to such appropriation.

Therefore, the actual reasons for the new role played by intellectual property in 1945 must lie elsewhere. This paper is an attempt to discover those reasons. We will see that the central prerequisite for the appropriation of intellectual property by the victors in 1945 lay in a quasi extra-legal constellation which for the first time removed the barriers to such a process which had traditionally existed. The prerequisite for this was the total victory of the Allies, which gave them complete authority over the losers.

This raises the second question: what was the legal character of the transfer of intellectual property? Was it a specific form of reparations? The thesis presented here is that the process cannot be subsumed under either the concept or practice of reparations as these have developed in the 20th century, nor was a separate new legal principle created for the purpose.

In the approach taken here, the issues of consequences and success must be ignored. It is irrelevant from a legal standpoint who ultimately profited most from the transfer of intellectual property. If the process turned out to be relatively harmless, or even favorable for the losers over the long term, this does not permit any conclusions regarding the intentions of those who planned and executed the process, nor does it change the legal character of the process in any way.

The service of specialists was certainly at least as important for technology transfer as appropriation of concrete knowledge. Nevertheless, the former is not taken into account, for reasons having to do with methodology. It is virtually impossible to determine the ratio of coercion to the free will which existed when German experts, in particular, worked for the victors.

I

To what extent can the processes investigated here be referred to as reparations? Is the term "intellectual reparations" appropriate?[3]

To answer this question, we must distinguish between a *de facto* concept of reparations and one based in law, specifically international law. In a *de facto* sense, reparations can be described as unilateral transfers of economic assets from the losers to the victors after a war.[4] This, too, requires a legal criterion, however; it is civil property deeds that are transferred, not sovereignty. On the other hand, no concern is given either to the legal foundation for the victor's claims or the reasons for the loser's obligations.

The concept of reparations under international law is more narrow. It assumes that the transfer is occurring on the basis of a contractual agreement between the victor and loser.[5] This agreement is generally a component of the peace treaty; however, it may also be part of a preliminary peace treaty, an armistice or a separate agreement. Formally, therefore, the transfer is voluntary, based on a contractual obligation, although the actual basis for the transfer, of course, is material inequality — for the debtor, the reparations are a consequence of losing the war. Still, it is a process governed by law. In particular, the victor is entitled only to that which the loser has granted him in a legally binding form. He cannot simply take what he wants. The loser alone is responsible for producing the reparations.

The obligations covered by the two concepts coincide when the victor has no direct authority over portions of the loser's property, i.e., cannot take anything that the loser is not willing to grant him. This is the case when the victor is not occupying any of the loser's territory when the war ends. The only exception to this are the foreign assets of the loser in areas under the control of the victor. However, if the victor is occupying portions of the loser's territory, then the *de facto* authority over the loser's assets is generally greater than the legal authority.

A special case exists when the loser at least temporarily loses his status as a subject of international law and is not immediately annexed. In this case, only the *de facto* concept of reparations remains applicable, since there is no contractual partner available for a reparations agreement.

Which of the two concepts is to be used depends on the circumstances. Neither alone can be regarded as adequate to the task. If one is interested in the amount of assets actually transferred, only the *de facto* concept is relevant. Otherwise, there would be a false light cast on a victor who appropriates all the assets he wants from the loser in advance of the reparations agreement and later, in a gesture of apparent generosity, renounces reparations in the peace treaty. On the other hand, if one's interest is in the legal foundation of the property transfers, then the legal concept should be used.

Before these questions are answered concretely for the situation of 1945, we need to look at the historical background of reparations and at the concept of reparations.

II

It is only since World War I that the expression "reparations" has been used in the sense which is prevalent today.[6] The matter to which it refers,[7] however, is probably as old as peace agreements between politically organized entities capable of protecting the property of their members. Two forms of property transfer can be distinguished: those involving sovereignty and those involving civil property deeds. The former are cessions of territory, the latter are reparations or, as they have traditionally been called, war indemnities.

Peace treaties rarely contain justifications of territorial concessions, but merely codify them. War indemnities are more often justified, although these are usually pseudo-justifications — the loser shall pay, for example, the victor a certain sum "for the purpose of covering a part of the expenses incurred in the war".[8] The actual justification is not stated, any more than it is in the case of territorial concessions, because it is always a given in both cases. It is based on the rights of the victor. Grotius describes this particularly drastically as having been originally completely unrestricted.[9] According to Grotius, under natural law, the victor wins complete authority over the loser, not only over all of his property, but over the life and liberty of all his members. Thus, all actions of the victor are legitimized on the basis of victors' rights. Grotius then attempts, as does the entire European international law tradition, to show the *de facto* evidence and normative justification for the fact that, in actual, applicable law, many restrictions have been implemented, particularly between Christian nations.[10] This applies primarily to the status of persons, but not to cession of territory or war indemnities, the right to which was not seriously challenged in modern Europe in either theory or practice until World War I. Nevertheless, it was always silently assumed that some sense of proportion would be maintained. This was important, above all, for war indemnities, and had to do not only — and not even primarily — with generosity or humanitarianism on the part of the victors, but more with the practical limits of the instrument. War indemnities are much less common in the European tradition than cessions of territory. This seems partially due to the fact that territorial gains were more attractive than financial ones for most victors. Probably more important, though, was the awareness of the limited resources of the loser, who had often exhausted all of his financial and economic options. If necessary, he might have been forced to pledge hefty war indemnities. In practice, however, he probably would not have delivered much. There is also a third factor. Cessions of territory were relatively easy to implement. In many cases, the new owner had already occupied the ceded territories during the war. Otherwise, they were transferred to him at the conclusion of the peace treaty at the latest. The surrender of war indemnities, however, depended much more on the will of the loser. Even if he was in possession of the assets he had promised, he still had to actually surrender them. This could be delayed in many ways. The last resort available to the victor was collection by force. However, long after the war and the conclusion of the peace, this was often unpopular and always expensive, and could easily misfire. For while an invading army, assuming it is strong enough, can gain control of a territory with no great difficulty, movable property can be relatively easily withdrawn from its grasp.

There were several options for implementation in theory and practice. The most important role by far was played by monetary payments and payments in kind. The big advantage of currency was that it afforded both sides maximum freedom. The creditor could buy whatever he wanted with it, and the debtor was free to determine how he would come up with the money. He could release reserves or sell foreign assets, intensify trade or increase production. There was one difficulty with this last option, however. The debtor had to sell his products in order to earn the money for the indemnity. This made him dependent on factors out of his control having to do with the world economy. In addition, the influx of large sums of money could cause inflation for the creditor.

Both dangers could be avoided by in kind payments, which also had the advantage for the debtor that his products were guaranteed delivery. The disadvantage was that he did not have flexibility in planning his production. The creditor, for his part, had to decide on certain goods in advance. As a result, currency was generally preferred, even in ancient times.

How and under what conditions would transfer of intellectual property have been possible in this context, and why did it not occur in practice? There would have been at least two difficulties.

1. If war indemnities are established between two former warring parties, they must be quantified in some way. Normally, a sum is established. It is also conceivable, however, that the victor would receive a certain percentage of the volume of trade, taxation income or national revenue. In any case, it is in the interest of both parties that it be more or less possible to assess the value of that which changes hands. This is not the case with the transfer of intellectual property, the value of

which it is impossible to calculate. The events occurring after 1945 demonstrate this.[11] Particularly characteristic in this regard is Gimbel's extraordinarily thorough and excellently documented study which, on the question of the value of reparations, purely speculatively assumed that Molotov may have been correct in his claim at the conference of Foreign Ministers in Moscow in 1947 that, up to that point, the Western powers had drawn from their zones reparations with a value of over $10 billion.[12] To date, the only assessment method to be taken seriously has been proposed by Buchheim.[13] It is based on development costs and is capable of providing valuable starting points. After all, patents also have a market value. Ultimately, however, the actual value of such intellectual property can only be determined after the fact. This is particularly true when a national economy is looked at as a whole. Of course, none of this excludes the possibility that a victor and loser will nevertheless agree on transfer of intellectual property, but such a process is afflicted with great uncertainty, especially for the victor.

2. Intellectual property can be easily hidden or even destroyed. It can never be seized as easily as material property or currency. This is true even when the debtor nation acts as the loyal executor of the creditor. If, on the other hand, the debtor state uses obstructive tactics like his members, which is more likely, then the creditor has essentially no chance of monitoring whether he has actually received the demanded property or whether he has been provided with only incomplete, outdated or "neutralized" documentation. Thus, a claim for intellectual property to be handed over as part of a war indemnity is risky for the victor/creditor unless he can collect this property himself with his own hands, and even then there are manifold possibilities for withholding important material from him.

III

What changed, other than the name, when the expression "war indemnities" was replaced with "reparations" at the end of World War I? In the light of the conditions from which it arose, there was a clear narrowing of the concept.[14] As a result of the intensity of the fighting and the bitterness of the conflict, many people related the demand for war indemnities in 1918–19 less to the ability of the loser to pay than to the total costs incurred by the victors. In contrast to this, the concept "reparations" had a considerably narrower range of application. It related not to the costs of war, but to amends for damage done, i.e., to compensation for damages and indemnification of losses suffered. Broad interpretation could still lead, however, to immense claims, so further restrictions were nearly inevitable. This situation was reflected in the Versailles Treaty, which first established a general obligation of Germany to provide reparations for all the loss and damage suffered by the Allies (Article 231), but then limited this to damage suffered by the civilian population (Article 232).

This change in favor of the loser, which was intended to protect him from excessive claims, had two problematic consequences. The discussions among the victors, instead of being pragmatic considerations of the loser's ability to pay, were dogmatic debates of what categories of damages should fall under the concept of reparations. It was only pseudo-protection of the loser anyway, since it was incorrectly implied that victors in earlier times had simply imposed on the loser the obligation to pay all the costs of the war, without regard to his ability to pay.

The second problematic consequence was the link which developed between the question of war guilt and the concept of reparations. War indemnities simply resulted from the victor's rights, whereas reparations were based on acts of injustice, and the right of the victors to reparations was ultimately derived from the placing of war guilt.[15] Whereas one had previously drawn this right from the victory, it was now an injustice theoretically unrelated to victory or defeat that formed the basis, although it was clear that victory would remain the essential condition for the possibility of claiming reparations.

The link with the question of guilt led to moralization of reparations. The debtor had to pay reparations because he was responsible for the war, not because he had lost it.

The transition from war indemnities to reparations thus essentially represented a shrinking of the rights of the victor, even though the actual material claims reached record highs. Reparations were supposed to serve to restore the conditions which had existed prior to the war. In the case of monetary payments, this correlation could be established only marginally, since money could be used for anything. It was different for deliveries of goods, for which it had to be recorded in one way or another what damage and losses caused by the loser the goods were intended to compensate.

In this regard, claims for transfer of intellectual property were difficult to justify. No matter how much damage the losers may have caused by the war for which, according to the peace treaty, they were responsible, they could hardly have taken notable amounts of intellectual property from the victors. In *this* respect, no damage had been done. It was even difficult to make the argument that it was indirect compensation, with the intellectual property serving as compensation of equivalent value for the losses suffered by the damaged parties. For those who would profit from the transfer of intellectual property were generally not the same people who had been damaged most by the war.

Thus, the concept of war indemnities permitted claims for intellectual property, but they were prevented by various practical factors, whereas the concept of reparations clearly excluded them, if one bases one's assessment at all on the content intended by the victors in 1919.

IV

The general understanding of reparations has not undergone any other fundamental changes in the 20th century, although the moralization increased as time went on. It had consequences in World War II. This time, the question of guilt was even more clear in the minds of the victors than in 1914–18. This resulted in almost a kind of compulsion to obtain reparations. A failure to do so could have been interpreted as proof that the enemy was less or not at all responsible for the war. Therefore, even before the war ended, the victorious parties determined that the losers would have to pay extensive reparations. There was never a fundamental debate among the Allies on the advantages and disadvantages of reparations, although the experience gained between the wars was far from universally positive. Even those nations went along who had little or no interest of their own in demanding reparations, or who would have preferred to use the resources of the losers for other purposes, such as repayment of pre-war debts.

Despite the fundamental similarity, there were considerable differences in practice between 1919 and 1945.

1. The reparations of the period between the wars were anything but a success.[16] They put heavy strains on international relations and, in the end, the creditors received much less than they had claimed and hoped for. Monetary payments had caused special problems. In order to obtain the necessary currency, Germany had to push its exports. The victors were major potential customers, but were not willing to open their markets to the necessary degree. As a result, the victors of World War II agreed early on that no monetary payments would be demanded this time. They stuck to this agreement into the early 1950s. In this way, the reparations which had been traditionally most important — even at Versailles — were lost. The victors therefore felt justified in, indeed virtually coerced into, establishing additional reparations categories which posed no risk to the world economy.

2. The strongest victor by far, the United States, was in a situation which made traditional reparations appear not only uninteresting but actually undesirable.[17] The country had suffered no direct damage from the war. Its economy had grown enormously, and overproduction rather than scarcity was anticipated for the post-war period. In kind delivery of goods by the losers would only mean increased competition at home or abroad. However, since it was firmly established for political reasons that the losers were to provide reparations, there was an incentive to define other reparations categories which were more attractive to the United States, and there existed some very clear ideas on this subject. Edwin B. Pauley, the US delegate to the Reparations Commission, stated it clearly on September 19, 1945: "We cannot use plants, machinery and labor. But we can take and should assert to the fullest extent our demand for gold currencies, foreign assets, patents, processes, technical know-how of every type."[18]

If one proceeds from the original understanding of reparations, then it is a questionable matter for intellectual property to be subsumed under the concept of reparations. At any rate, the character of compensation for damages was not immediately evident, at least from the US perspective. The situation looked a bit different from the point of view of the continental European nations. The Germans had indeed seized intellectual property in the countries they had occupied, particularly in France. In these cases, however, as with the return of stolen works of art, amends were to be made under the title of restitution, which was an undisputed part of international law, associated, if necessary, with specific compensation. One could hardly derive from this a legal entitlement, for example, of the US to take any German intellectual property it desired.

3. The original intention of reparations — to obtain as large a contribution as possible from the losers as restitution for the damages and losses suffered by the victors — now competed with another viewpoint. It was understandable that, after the bloodiest and most abhorrent war in the history of the world, there was a desire to permanently weaken the aggressor. However, this objective was bound to collide with that of maximizing reparations. If one wanted to get as much as possible out of the loser, he had to remain economically viable. If one wanted to be safe from him, then his economy had to be destroyed. One could not have both at the same time. At best, a compromise might be achieved. The form in which it presented itself was that the victors would appropriate as much as possible of the loser's economic substance. Dismantling became the classic example of this process, but it also included the complete confiscation of foreign assets and appropriation of as much intellectual property as possible.

V

The ruling out of monetary payments thus led the victors to establish various reparations categories which were previously uncommon or completely unheard of. This shift had important consequences. Traditional war indemnities, and even lump-sum reparations claims, particularly if they were to be paid in currency, had given the debtor greater flexibility than the special payments and measures now being demanded. Moreover, these were more in the nature of punishment or of weakening the loser's potential by reducing substance than was the case in 1919, and much more so than in the case of war indemnities. Of all these measures, only the seizure of foreign assets had also occurred after World War I.

The new reparations categories not only had a more pronounced punitive character, they also required a significantly stronger enforcement mechanism and, in this respect, did not help improve relations between the victors and losers. Procuring currency not only gives a government greater flexibility, it is also less threatening to its legitimacy than other demands. If a government is forced to dismantle or simply destroy extensive production facilities, or if it must systematically comb its economy for exploitable intellectual property, it is very likely to have legitimation and enforcement difficulties. The affected parties will try to withhold as much as possible from the government, but the government will probably join the resistance.

These difficulties were hardly present in 1945, however, due to the almost unrestricted power of the victors. Not the least of their reasons for establishing this new type of reparations claim, we may assume, was that they would not have to worry much about enforcement. With the exception of Finland, all loser nations were occupied by Allied troops. Although normal reparations agreements were signed with all of the losers except Germany and Japan, the occupation of territory also permitted the victors to take by other means whatever they laid claim to. The degree to which they made use of these mechanisms varied, but overall it was extensive. In Germany, the unconditional surrender even included transfer of governmental power to the victors, who responded according to their preferences. While the Soviets and French, who had been very adversely affected by the war, concentrated primarily on traditional categories, especially deliveries from current production, the Americans stayed primarily with foreign assets and intellectual property, and used dismantling predominantly as a means to weaken the enemy rather than for their own enrichment.

VI

The victors thus had nearly unlimited power to appropriate the assets of the losers. However, the concept of reparations generally implies the idea of and right to a legal process. Therefore, the question which remains to be asked regards the legal basis for the confiscation of intellectual property after 1945. The distinction between the *de facto* and legal concepts of reparations outlined at the outset shall serve as a starting point.

Only the reparations from the smaller loser nations agreed upon in the amistices of 1944/45 and the peace treaties of 1947 could be considered contractual reparations.[19] All of the different types of German and Japanese reparations up to the 1950s, on the other hand, corresponded to the *de facto* concept, as did the many additional appropriations from the smaller nations. Therefore, we cannot speak of normal reparations in accordance with international law.

As clear as this conclusion is, it is also unsatisfactory. It would be senseless to argue that, because the loser had no government, all appropriations undertaken by the victors, whether under the designation of "reparations" or not, were *ipso facto* unlawful. Under such an argument, the loser would gain enormous advantage from total defeat and unconditional surrender in the form of a guarantee of his property. A "normal" loser would never be given such a guarantee but would have to expect reparations to be imposed on him in the context of the peace treaty. Therefore, other criteria are presented here for determining the legal status of these processes.

1. Since the victors had no contact person whatsoever for purposes of negotiating reparations, at least not in Germany, an analogy is necessary. In such a situation, a victor with unilateral authority over the property of the loser has no more and no fewer rights than he is normally capable of obtaining contractually. He is entitled to take from the loser those categories of reparations which have been established contractually in other places or in earlier cases. The reparations negotiated after World War II were amazingly restrictive in this regard. They were limited essentially to deliveries of goods and confiscation of foreign assets.[20] However, one also has to at least take into consideration the precedent

of World War I. Then, too, however, there had been a pronounced concentration on payment in goods and currency, in addition to the seizure of foreign assets.

If one accepts this argumentation, then post-1945 dismantling (as reparations, as opposed to demilitarization measures) and, particularly, the confiscation of intellectual property, are not supported by the relevant practice to date under international law.

2. However, one may go a step further and ask to what extent such a process can be derived from the generally accepted concept of reparations, the basis of which was responsibility for the war, which could be used to justify punitive measures. Any measure which causes damage to the guilty party can be declared punishment. If one interprets reparations in this way, then its legal basis is reduced to the old victor's rights, according to which the victor has unlimited authority over the loser. However, this kind of relapse into the natural state was never intended when the concept of reparations was originated at the end of World War I. Instead, reparations were supposed to serve as compensation for damages. As outlined above, however, confiscation of intellectual property was difficult to reconcile with this definition. Therefore, from a legal viewpoint, it could not be derived either from precedents or from the general principles of the reparations concept, which were undisputed even in 1945. Thus, from this perspective, it cannot be regarded as having a legal foundation but must be viewed as an arbitrary measure.

The matter of definition is not just an intellectual game after the fact. The victorious powers took it entirely seriously and confirmed the interpretation presented here. In the Potsdam Protocol, they agreed among themselves that the reparations claims of all victorious nations except the Soviet Union and Poland would be satisfied from the western zones of Germany. To this end, the assets collected would be pooled and distributed according to a specific key.[21] On December 21, 1945, the Inter-Allied Reparations Agency (IARA) was established to carry this out. The most important objects it was to distribute were the dismantled plants.[22] From the perspective of the losers, the primary effect of the dismantling was frequently destructive, especially when little of that which was dismantled could be reused. This perspective of destruction, or industrial disarmament as it was called, also played a big role on the Allied side; it even had clear priority in the American conception of dismantling. Internally, in the US, various parties argued that dismantling must proceed even if no-one could be found who could use the facilities, or if they were not reusable, which amounted to destruction.[23] Nevertheless, or perhaps precisely for this reason, the reluctant German authorities were prodded by the British and Americans to designate the dismantling as reparations.[24] Thus, not only was it treated as reparations by being placed under the jurisdiction of the IARA, but the corresponding terminology was also imposed, despite the fact that not even the *de facto* concept of reparations could be applied to the cases of pure destruction, since no assets were transferred.

In terms of reuse and transfer of assets, intellectual property was certainly less problematic than dismantling. Yet the Western occupation powers did not place it under the jurisdiction of the IARA — a fact of which, curiously enough, there has never been any notice taken. This can only lead to the conclusion that these appropriations were not officially considered reparations. Otherwise, they would have been an unlawful appropriation at the expense of the IARA partners. This finding holds despite the fact that there was no strict usage of terminology. Internally, and even in press releases, the term "intellectual reparations" was occasionally used,[25] but in actuality the matter was not handled as a reparations process.

The motivation for this willful system is obvious. The advantages to be gained were so great that there was no desire to share them with the smaller victor nations. The dismantled plants, meanwhile, were of less interest to Great Britain and no interest whatsoever to the US, so they were quite happy to give more generously in this regard. The process shows, however, that the victors were aware of the problems with their procedure. There was even a kind of unconscious retreat to the victor's rights. As an official put it before a Congress committee in 1947: "The fundamental justification of this activity is that we won the war and the Germans did not."[26]

If no legal basis for the confiscation of intellectual property can be derived from the traditional and generally accepted concept of reparations, then one must ask whether it could be derived in some other way from the position of the victors in defeated Germany. This leads to the question of the legal status of Germany after World War II and thus to one of the most complicated and persistently controversial problems of international law.[27] We shall only address the consequences arising from different perspectives on the question of confiscating intellectual property. This will be done by highlighting two extremes from among the many proposed interpretations.

1. Although hostilities ended on May 8, 1945, the state of war persisted. Germany was occupied. Therefore, one could speak of a belligerent occupation, an *occupatio bellica*.[28] This is a condition with a rather precise legal definition for which, in particular, the Hague Land Warfare Convention of 1907 undertook extensive codification of the rights and obligations of the occupation force,[29] which is by no means granted unlimited authority. Private property, in particular, enjoys extensive protection (Article 46). Enemy property may be confiscated only for specific purposes and only in exchange for compensation or at least a promise of future compensation (Articles 51–53). To be sure, there is a proviso concerning necessities related to the conduct of war, in the interest of which the occupation force may appropriate practically anything (cf. Article 23). However, after Germany's capitulation, or at the latest after that of Japan, no further war was being conducted, and the proviso was moot. A second proviso was significantly more limited. It stated that property could be appropriated to meet the needs of the occupation (Articles 49, 52). It was applied very extensively in the first years by all of the victors under the heading of occupation costs. This means, conversely, that, if one assumes a situation of *occupatio bellica*, then the appropriation of private property by the victors which was declared to be reparations or their equivalent was without legal foundation to the extent that it was not derived from the needs of the occupation. Where the boundary is to be drawn in this respect is disputable.[30] It seems indisputable, on the other hand, that not all appropriations undertaken can be justified in this manner. For example, the dismantling

was hardly necessary for the occupation. Under some circumstances, though, it could be declared disarmament measures. Appropriation of intellectual property, on the other hand, neither served the purpose of disarmament nor could it be derived from the needs of the occupation.[31]

2. However, one could interpret the unconditional surrender entirely differently. Inasmuch as the Allies had conquered Germany and taken over supreme authority, Germany had been incorporated into the territory of the victor and one could speak of annexation. Thus, reparations became a domestic matter of the victors. Each nation had the right to appropriate property, either in return for compensation or as punishment. One could interpret the measures against Germany as punishment, which would explain their unilateral nature (since property was not appropriated in the rest of the victors' territory). The problem with this is obvious. It would have to have been arbitrary and indiscriminate collective punishment, for which there was no basis in the domestic law of the victors. In addition, it is questionable to speak of annexation, particularly since the victors had expressly denied any intent of annexation.[32]

There are many more possible interpretations of the legal status of Germany after World War II between these two extremes, but they cannot be addressed here. Still, no perspective has yet been advanced which would permit us to derive a legal basis for the confiscation of intellectual property. It would also be difficult to explain how property which is widely protected even under the strict conditions of wartime occupation could be up for grabs during peacetime.

Therefore, the right to confiscate intellectual property cannot be derived from the generally accepted concept of reparations, nor from the position gained by the victors as a result of Germany's unconditional surrender, nor from any other measure supported by the practice of international law. This does not mean, on the other hand, that such confiscation is inherently unlawful for all time, that it is fundamentally less legitimate than other measures of a victor. The legal basis for it simply did not exist in 1945. If this deficiency is to be eliminated, there must be further development of international law by either contractually agreeing to these kinds of measures or at least modifying the general concept of reparations. However, there was not the least attempt to do so in 1945. Unlike in the area of war crimes, for example, the victors did not use their absolute power to change existing law governing matters of reparations.

Translated by Sally Robertson

NOTES

1 The term "intellectual property" generally tends to be used in legal terminology for copyright matters, i.e., for individual legal titles to literary and artistic, but also scientific creations, rather than for the multi-faceted, technical and scientific knowledge and resulting processes which form one of

the essential foundations of a modern national economy. There is no generally recognized umbrella term for this. In legal dogma, patent law is the most important example. The phenomena addressed here, however, deal with far more than just patents. For lack of a more precise expression, the very broad (basically somewhat too broad) term "intellectual property" will be used.

2 This and subsequent general statements regarding peace treaties are supported by the sources consulted in Jörg Fisch, *Krieg und Frieden im Friedensvertrag. Eine universalgeschichtliche Studie über Grundlagen und Formelemente des Friedensschlusses* (Stuttgart: Klett-Cotta, 1979). This at least covers the European tradition reasonably comprehensively.

3 Cf. the cautious and distancing usage in John Gimbel, *Science, Technology, and Reparations. Exploitation and plunder in postwar Germany* (Stanford: Stanford University Press, 1990) VII and *passim* (Index on page 275).

4 For more detail, see Jörg Fisch, *Reparationen nach dem Zweiten Weltkrieg* (Munich: C.H. Beck, 1992), 17–18.

5 Cf., for example, Burkhardt Röper, "Reparationen". *Handwörterbuch der Sozialwissenschaften* (Stuttgart: Fischer, 1964) 8: 812–814; Ernst Féaux de la Croix, "Die Abgeltung der Reparationsschäden," *Die öffentliche Verwaltung* (1962) *15*: 212–213.

6 Cf. Fisch, *Reparationen* (see Note 4) 19–25, for additional information; also Bruce Kent, *The spoils of war. The politics, economics and diplomacy of reparations 1918–1932* (Oxford: Clarendon Press, 1989); Robert E. Bunselmeyer, *The cost of the war 1914–1919. British economic war aims and the origins of reparation* (Hamden, Conn.: Archon Books 1975); Marc Trachtenberg, *Reparation in world politics. France and European economic diplomacy, 1916–1923* (New York: Columbia University Press, 1980) and Fritz Dickmann, *Die Kriegsschuldfrage auf der Friedenskonferenz von Paris 1919* (Munich: Oldenbourg, 1964). First appeared in: *Historische Zeitschrift* (1963) *197*: 1–101.

7 No history of war indemnities has been written. Cf., meanwhile, Frieda Busch, *Tribute und ihre Wirkungen untersucht am Beispiel der französischen Zahlungen nach dem Krieg 1870/71*, (Ph.D. Thesis Basel 1931, Basel 1936); Nicholas Balabkins, *West German reparations to Israel* (New Brunswick, N.J.: Rutgers University Press 1971), 19–22, and *Corpus pacificationum. Systematische Zusammenstellung der Texte der Friedensverträge 1792–1913* (Berlin: Reichsdruckerei 1924).

8 Peace Treaty of August 13, 1866 between Prussia and Württemberg, Article 2. In: Clive Parry, ed., *The Consolidated Treaty Series*, Vol. 133 (New York: Oceana, 1969), [25]. For general information on the question of codification and justification, see Fisch, *Krieg und Frieden* (see Note 2), Chapter 4.

9 Hugo Grotius, *De jure belli ac pacis libri tres: in quibus jus naturae & gentium, item juris publici praecipua explicantur*, (Latin/English edition) Book 3, Chapters 4–9. (Oxford: Clarendon Press, 1925).

10 *Ibid.* Chapters 10–16.
11 For more detailed information, see Fisch, *Reparations* (see Note 4), 213–215.
12 Gimbel (see Note 3), 152, 170.
13 Christoph Buchheim, *Die Wiedereingliederung Westdeutschlands in die Weltwirtschaft 1945–1958* (Munich: Oldenbourg, 1990), 86–88.
14 See Note 6.
15 Cf. Fisch, *Krieg und Frieden* (see Note 1), 204–231.
16 Cf. Fisch, *Reparationen* (see Note 4), 25–29, for further references.
17 Cf. Friedrich Jerchow, *Deutschland in der Weltwirtschaft 1944–1947. Allierte Deutschland-und Reparationspolitik und die Anfänge der westdeutschen Außenwirtschaft* (Düsseldorf: Droste 1978), 19 ff.
18 Pauley to Byrnes, *Foreign Relations of the United States (FRUS), Potsdam Papers* 2, 511, cf. Pauley's instructions of May 18, 1945. *FRUS* 1945/3, 1222–1227.
19 Italy: Peace Treaty, Article 74. Romania: Armistice, Article 22; Peace Treaty, Article 22. Finland: Armistice, Article 11 and Annex H; Peace Treaty, Article 23. Bulgaria: Peace Treaty, Article 21. Hungary: Armistice, Article 12; Peace Treaty, Article 23. For texts of the amistices see Georg Friedrich von Martens, *Nouveau recueil général de traités. Troisième série. Volume 41* (Aalen: Scientia, 1969), 874–913. For texts of the peace treaties see *United Nations Treaty Series*, Volumes 41–42, 48–49.
20 See Note 19. For Germany and Japan, cf. Fisch, *Reparationen* (see Note 4), Chapter 3.
21 Potsdam Protocol Article III. *FRUS* Potsdam 2, 1485–1487.
22 Inter-Allied Reparations Agency. Report of the Assembly of the Inter-Allied Reparations Agency to its member governments (Brussels, 1951).
23 Cf. Fisch, *Reparationen* (see Note 4), 49 f., 244.
24 Hein Hoebink, "Demontage in Nordrhein-Westfalen 1947–1950." *Westfälische Forschungen* (1980) *30*: 56.
25 Gimbel (see Note 3), 28, 33, 52, 102, 175.
26 *Ibid.*, 174–175.
27 For an overview, see Rudolf Schuster, *Deutschlands staatliche Existenz im Widerstreit politischer und rechtlicher Gesichtspunkte 1945–1963* (Munich, 1963). Also, Georg Ress, "Germany, legal status after World War II," *Encyclopedia of Public International Law (EPIL)* (Amsterdam: North Holland, 1987) *10*: 191–206. A comparative analysis of all loser nations appears in Nisuke Ando, *Surrender, occupation and private property in international law. An evolution of US practice in Japan* (Oxford: Oxford University Press, 1991).
28 Cf. Michael Bothe, "Occupation, belligerent," *EPIL* (1982) *4*: 64–67.
29 October 18, 1907. Text in *The Consolidated Treaty Series* (1980) *205*: 277–298, for example.
30 For an analysis which is very much in favor of the occupied, see Hermann M. Maschke, *Das Krupp-Urteil und das Problem der "Plünderung"*

(Göttingen; Musterschmidt, 1951). Cf., also, William Gerald Downey, Jr., "Captured enemy property: booty of war and seized enemy property." *American Journal of International Law* (1950) *44*: 488–504. For general information: Ignaz Seidl-Hohenveldern, "Enemy Property," *EPIL* (1982) *3*: 167–170.

31 Cf. Downey as well as Yoram Dinstein, "Booty in land warfare," *EPIL* (1982) *3*: 66–68.

32 Preamble to the four powers' Declaration Regarding the Defeat of Germany and Assumption of Supreme Authority on June 5, 1945. Ingo von Münch (ed.), *Dokumente des geteilten Deutschland* (Stuttgart: Kröner 1968), 20.

CHAPTER 3

Exploitation by Integration? The Re-Orientation of the Two German Economies After 1945. The Impact of Scientific and Production Controls

Matthias Judt

The literature on the problem of Germany's reparations and the relationships between Germany and the Allies is vast. For some time, the topic was so widely discussed and debated, at least in the area of material reparations, that new findings were not to be expected. Before 1989, moreover, it was impossible to verify Western estimates on the reparations East Germany made to the Soviet Union between 1945 and 1953. In turn, East German studies, such as the one by Horst Barthel,[1] suffered under restrictions because researchers either had no access to archival records or the archives denied the very existence of files on reparations.

Restricted access to archival resources disappeared with the collapse of communism in Eastern Europe. This eventually enabled historians to renew discussion of the topic. Furthermore, John Gimbel's book on what he called "intellectual reparations" focused on a new aspect that dealt with deliveries and services underrated in the past, i.e., the confiscation of documents, blueprints, trademarks and patents, and the hiring of German scientists by Allied powers.[2] More recently, a book by Ulrich Albrecht, Andreas Heinemann-Grüder, and Arend Wellmann on the employment of German scientists and technicians in the Soviet Union after 1945[3] supplemented Gimbel's path-breaking study. Further studies, such as those by Tom Bower[4] and Linda Hunt,[5] concentrated on the conflict between the aims of denazification and Allied interest in getting at German scientific knowledge or potential, even if the people involved had a Nazi past.

Some of these studies, written from the point of view of political historians, underestimated the impact of Allied policies on the future economic development of the two German states. Moreover, some historians denied the existence of any competitive economic motives in Allied policies. Josef Foschepoth argued that such assertions misrepresented what actually happened. More likely, the Allies would have been interested in relieving the financial burden of occupation.[6] Although the cost of occupation had become a topic of political discussion, especially in the United States but also in France and Britain, Allied authorities indeed used the argument of competitive economic interest in their decision making. In this way, already in 1947 a "History of Field Information Agency, Technical (USA)" characterized World War II as a war that "became a war not only between armies but between nations, including their entire resources, scientific and industrial establishments, and civilian personnel."[7]

Wars generally grant the victors unique possibilities to determine the future economic development of the defeated country, above all by exploiting its industrial and scientific potential. This freedom to determine policy includes a choice of methods to be applied by the victor(s).

In the case of Germany after World War II, the first phase of Allied occupation policy was dominated more or less by "destructive" methods. The exploitation of existing German economic potential began with periodic dismantling (in the Soviet Zone), the apparently indiscriminate confiscation of documents, blueprints, patents and trademarks, the capture of German specialists, and the imposition of restrictive regulations and limitations on the future development of Germany's research and production. The latter set of actions applied to Germany as a whole. These destructive methods dominated Allied policy toward Germany until 1947–48.

The second phase began in 1948. The integration of the two German economies into newly established alliances, respectively, included the lifting of several bans on industrial production and scientific research as well as the active promotion of certain developments. The application of these "productive" methods enabled both eastern and western Germany to retain large portions of their scientific and production potential.

Although a periodic differentiation in the application of both kinds of methods is evident, we find elements characteristic of one period also applicable to the other. One of the historical ironies of this period is that the future division of the country was not as obvious during the "destructive period" as it would become during the "productive period." Furthermore, both policies contained inherent potential advantages for Germany, at least for those Germans in position to benefit from them.

1. The application of destructive methods

During the World War II, Allied preparations for the treatment of postwar Germany contained mainly proposals aimed at a massive reduction of Germany's economic and military potential. The destruction caused by Germany throughout Europe by the use of this potential was a justification for the United States, France, Great Britain, and the Soviet Union, which had suffered the greatest amount of physical damage, to establish a policy of treating Germany as a dangerous enemy nation. The Allies agreed in principle on the following goals:

1. Germany should make reparations in form of payments, the delivery of dismantled machinery, and the surrender of German assets abroad.
2. The Allies should establish controls over Germany's industrial production and its scientific research, i.e., the prohibition of all war-related work and the limitation of other kinds of scientific work and industrial production.
3. Moreover, the denazification of German society should become a major political aim of the Allied policy toward Germany.

Hungarian-born economic historian Yevgeny Varga's advice to the Soviet Union was the critical determinant in Soviet policy toward their zone. Already in October 1943, he proposed that Germany pay reparations to the Soviet Union in the amount of 15 billion Reichsmarks per year. Varga estimated that Germany had spent the same amount of money on armaments each year between 1933 and 1938. He therefore argued that the same amount should be paid out in reparations.[8]

The generally hostile climate toward Germany prevailing in the Soviet Union, whose population sought revenge for the damage wrought by the German *Wehrmacht*, obviously influenced the outcome of Varga's proposals. Since July 1941, or shortly after the start of the armed struggle between German and Soviet troops, propaganda material, e.g., Red Army newspapers, contained slogans such as "Kill the German!" This strong antipathy toward German soldiers only increased under the impact of the war, which Soviet soldiers witnessed personally in their westward counteroffensive.[9]

Furthermore, Soviet troops found letters and military orders in the personal belongings of dead German soldiers, in which either relatives of the soldiers demanded or superiors ordered the extermination of parts of the Russian population.[10] Using this evidence in the ideological preparation of their own troops, the Soviet military leadership created a climate among its forces that presupposed a hostile attitude toward Germany, which to some extent even tolerated crimes against the German civilian population.

Anti-German attitudes also existed among the Western Allies. As is well known, the Morgenthau Plan aimed at what was called the industrial disarmament of Germany and provided (included in Soviet proposals as well) for the use of German workforce abroad. Morgenthau proposed strong restrictions on Germany's industrial production generally.[11]

Although the Morgenthau Plan was not officially adopted, it indirectly guided American policy toward postwar German economic development. In March 1944, Secretary of State Riddleberger pointed out in a memorandum to President Roosevelt that Germany should be rendered incapable of waging war, that Germany's economic domination of Europe should be eliminated, and furthermore, that Germany should be made dependent upon world markets. These proposals included extensive controls over industry and foreign trade during the period of demilitarization and reparation, followed by a period of supervision of German industry. "Through lack of luxuries we [the victors] may teach them [the Germans] that war does not pay."[12] The undertone of orders for the occupational forces showed consistently the same kind of intended treatment. Joint chiefs of staff (JCS) directive number 1067 ordered the commander-in-chief of United States occupational forces to bring about "the industrial disarmament and demilitarization of Germany, with continuing control over Germany's capacity to make war."[13] Military Government officials should take no steps that "(a) [were] looking for the economic rehabilitation of Germany or (b) [were] designed to maintain or strengthen the German economy."[14]

British authorities also discussed extensive restrictions on scientific research and university training. A paper of the intelligence corps proposed in 1945 that all universities and the polytechnics in the British Zone, except Göttingen University, should remain closed. In addition, the paper contained suggestions to reduce Germany's industrial research by abolishing pure research and development altogether or limiting it to a minimum.[15]

Postwar investigations into the structure of the German economy confirmed expectations that many had anticipated during war time. Germany's economy clearly had been organized on the basis of a war economy. Whereas the total output of Germany's industry had grown by 16 per cent between 1936 and 1939 and declined to 85 per cent of 1936's output in 1944,[16] the production of producer goods had doubled and the production of armament goods was 12 times higher in 1944 than in 1936.[17] This was made possible by two measures. First, large investments in producer good's industries between 1933 and 1939 increased their share in total investments from 55.5 per cent in 1933 to 81.1 per cent in 1939.[18] Second, changes within the employment structure in the economy as a whole converted it into a war economy. In November 1943, most of the employees in several branches of German industry were working in fulfillment of orders in the interest of *Wehrmacht* or armaments production. The structural changes within industrial branches with peaceful uses were strongest in the founding industry (68% of the employees worked for *Wehrmacht* and armament orders), the engineering and automotive industry (75.1%) and the sheet-metal industry (82%).[19]

Given such preconditions Germans could expect the Allies to move quickly and forcefully to dismantle this war economy. In the first months of occupation, all of the Allies looked for possibilities to enforce new structural changes within the economy and the economy's general restriction to "non-dangerous" production. This meant that even industries with peaceful uses were to be restricted. In February 1945, the leaders of the Soviet Union, Great Britain, and the United States agreed at Yalta that after the start of Germany's occupation the three powers would "eliminate or control all German industry that could be used for military production" and that they would force Germany to "exact reparation in kind for the destruction wrought by the Germans."[20]

Measures taken by the Allies between May and August 1945 resulted from the fear that Germany's scientific or military potential could merge with the potential of the one Axis power still fighting, namely, Japan. This policy was, however, not changed after the latter's capitulation. More important, until 1946 it was impossible to reach agreement on regulations concerning German industry and scientific research. Such regulations included the confiscation of material and immaterial German property; the partial dismantling of industrial and scientific facilities; prohibitions or, at least, limitations on German research and production; and the capture or hiring of German "specialists" under privileged terms or under pressure brought to bear on them.

The establishment of occupational authority in Germany by the Allies also initiated Allied control of industrial and scientific facilities. Early assessments of Germany's industrial and scientific capacity enabled units of military personnel, including individuals mobilized from the private sector, to determine whether or not these facilities should be removed to Allied countries or destroyed. In addition, the Chiefs of the Occupational Forces issued specific basic agreements among all Allies on the control of research and development in industry and science, work that was continued after the war by the Allied Control Council. These included the so-called "Additional Claims," agreed by the Chiefs of the Occupational Forces at Soviet Headquarters in Berlin-Karlshorst on June 5, 1945, and published as "Proclamation No. 2" of the Allied Control Council,[21] the "Level-of-Industry Plan" of March 29, 1946, and the Laws of the Allied Control Council Nos. 25 and 29, issued on April 29 and May 31, 1946, respectively.

All these regulations, but especially the Level-of-Industry Plan, aimed to limit Germany's industrial production until 1949 to 63 per cent of what it had been in 1936. They even imposed heavier limitations on the iron industry (to 31%), the automotive and aircraft industries (to 30%), and the metal industry (to 39%). Had this plan been fully realized, the result would have meant that, by 1949, 28 per cent of the workforce as of 1936 would have been unemployed.[22]

To reach such low production targets it was necessary to dismantle a great deal of existing industrial plant. At least in eastern Germany dismantling of this magnitude had, in fact, taken place since the end of the war. In Saxony, the most industrialized area of the Soviet Occupation Zone, the first stage of dismantling took place between July and October 1945. As reported by the Economic Chamber of Dresden/Saxony (*Wirtschaftskammer Dresden/Sachsen*), most of this activity occurred through August, followed by a lull until the end of October.[23]

Similar reports on dismantling in other eastern German states (*Länder*) are also available.[24] Quite obviously, Soviet local and regional commanders wanted to satisfy potential claims their country had on German plant and equipment, even if an agreement on the total amount of reparations had not yet been fixed.[25]

Whereas the dismantling of facilities was concentrated in eastern Germany, regulations on scientific control were enforced equally in each of the four zones of occupation. Allied Control Council (ACC) Law No. 25 restricted or even prohibited scientific research by German institutions. On the surface these regulations focused on scientific research with militarily application. The possible dual application of some of this research, however, limited or prohibited further developments even in some civilian areas. The law's "Schedule 'A': Prohibited Applied Scientific Research" included aerodynamics, aeronautical structural engineering, aircraft power plants, hydrodynamics, and ship construction. "Schedule 'B': Applied Scientific Research Requiring Prior Permission" included, for example, electromagnetic, infrared, and acoustic radiation research, the purpose of which would be to facilitate "communication of intelligence by telephone and telegraphy, the provision of public broadcast or television services."[26]

The enumeration of specific technologies at the high end of German scientific and industrial research threatened the country's scientific program as a whole. The entire Allied approach seems to suggest that the Allies intended to threaten the possibility for Germany to recover economically as well.

Furthermore, a subsequent ACC law, No. 29, ordered that "any of the four powers in occupation of Germany ... may request in writing an authenticated copy of any book, paper, statement, record, account, writing or other document from the files of any German industrial, business, or commercial enterprise."[27] This law enabled the Allies to confiscate virtually every document, regardless of whether or not it belonged to the state or to private individuals.

The Americans in particular took advantage of Law No. 29. Branches of the Field Information Agency, Technical, USA (FIAT), processed over 29,000 reports, confiscated 55 tons of documents, and made over 3,400 trips within Germany to investigate the so-called "targets" of interest through June 30, 1946.[28] Considering the transatlantic transportation problems they were already experiencing, the Americans preferred to seize "lightweight" goods, such as documents, as compensation for their participation in World War II.

On the other hand, the Soviet Military Administration looked, as previously shown, for machinery and equipment. It is obvious that, at the beginning, they underestimated the value of patents, trademarks, and blueprints. They even left behind the library of the *Reichspatentamt* in Berlin, which the Americans confiscated then when they arrived in mid-1945.

Belatedly recognizing the value of those documents, the Soviets were forced to suggest additional regulations to ACC authorities. At the end of 1945 they proposed to the Reparations, Deliveries and Restitution Directorate of the Allied Control Authority that in "the case of absence of technical documents, necessary for the assembling and exploitation of the [dismantled] equipment at the new place, the Zone Commander ... [was allowed to] order German specialists working in the plant to prepare all necessary blueprints and documents as soon as possible."[29] The other delegates in the Reparations, Deliveries and Restitution Directorate of the Allied Control Authority immediately agreed.[30]

This train of thought has to be completed by the explanation of the "advantages" Germany could have even under those conditions. Although dismantling and confiscations of German property of any kind caused uncertainty as to the country's own recovery, measures had only short-term effect. Moreover, the dismantling of facilities that had been necessary in the German war economy, made it unnecessary for German companies to close down such facilities themselves. The dismantling of other facilities in western Germany created pressure to replace these losses within a few years.

On the other hand, the Allies, especially the Soviet Union and France, knew that they could get only used machinery. This gave rise to a willingness to change reparation policy. The Soviets and the French started to look for reparations out of production instead of — or at least in addition to — reparations in the form of dismantled factories. In the case of eastern Germany this change in reparation policy produced the economic demand necessary for the country's recovery. In western Germany, British opposition to the extraction of reparations out of production rendered a similar procedure impracticable.

2. The application of productive methods

Although all-Allied restrictions on production and research effected both areas, the Soviet and the Western zones, in much the same way, western Germany drew some advantage from the postponement of some reparations until the 1950s. According to Fisch's calculations, between 1945 and 1953 western Germany spent approximately $16.8 billion[31] on reparations and costs of occupation but got credits to the amount of $1.5 billion in return during the same time.[32] According to the London Agreement on Debts, after 1953 (until 1988) it spent an additional $15.4 billion for public and individual reparations, occupational costs, and services.[33]

The estimates for eastern Germany, partly confirmed by new archival material, vary from the official figure of about $4.3 billion given by the Soviet Union in 1953. The Soviet announcement did not include services like occupational costs, "hidden reparations," money confiscated by the Soviets (*Besatzungsgeld*), and illegal confiscations.[34] Therefore, the actual figure must be at least approximately $13.3 billion. Vague estimates of *Besatzungsgeld*, losses in eastern Germany's foreign trade, and the costs of transportation of reparation goods,[35] on the one hand, and the item of hidden reparations, on the other, would increase the total amount to about $15.3 billion.

The Soviet Union[36] recognized, in part, the following items as reparation:

1. Dismantling: $2.436 billion[37] (recognized: $1,484 billion);
2. Deliveries out of current production: $2.6143 billion[38] (recognized: $2.473 billion);
3. Deliveries of the Wismut SAG: $1.5845 billion[39] (recognized: $335 million).

On the other hand, the USSR did not recognize any of the items, listed below:

1. Occupational costs: $5.9141 billion;[40]
2. Repurchase of SAG companies: $382 million;[41]

3. Illegal confiscations: \$352.1 million;[42]
4. Estimate about *Besatzungsgeld*[43]: \$1.24 billion;[44]
5. Losses in East Germany's foreign trade: \$400 million;[45]
6. Costs of transport: \$133.3 million;[46]
7. Hidden reparations: \$266.7 million.[47]

If we substract the expenditures for the costs of occupation, the calculation shows that eastern Germany fulfilled 94.1 per cent of the Soviet demand for \$10 billion in reparations. Therefore, it is no wonder that the economic recovery in the Western zones was more successful than in the Soviet Zone. The Western zones managed to reach the production level of 1936 earlier than the Soviet Zone (West Germany: at the end of 1949;[48] East Germany: at the beginning of 1951[49]).

The basis for economic recovery in both Germanies was different as well. Whereas East Germany concentrated on the development of its mining, steel, and parts of the chemical industries, West Germany based its recovery on the production of energy, building materials, fertilizers, textile, and leather products.

The Soviet Union focused its interest on deliveries from industries that had been, from the standpoint of the Western Allies, the basis for the German potential to conduct war. That is why Soviet authorities encouraged East German authorities to redevelop precisely those industries restricted in the West. On the other hand, the ideology of "socialist industrialization", which granted priority to heavy industry, limited the development of consumer good's industries in the East. In West Germany, those industries were able, earlier than producer goods industries, to increase their gross fixed capital.[50]

Soviet corporations (*Sowjetische Aktiengesellschaften* or SAG) played an important role in East Germany's recovery, one based on heavy industry. They were a unique kind of economic structure because their organization operated under the rules of bourgeois civil law in a country that was on its way to becoming a communist society. With the foundation of SAG companies, Soviet authorities were able to claim that plants, especially those that had been necessary for Germany's war economy, were "out of German control." Accordingly, they were also able to claim that the power of German monopolies had been broken. On the other hand, the Western Allies understood the act of breaking up monopolies as the process of decartelization, and, from the standpoint of the West, the SAG's were new cartels, although this time with different owners.

The existence of SAGs gave the USSR the possibility to influence directly East Germany's economic development, since the Soviet Union was in a position to control large portions of East German industry for its own benefit. Between 1947 and 1951, the average percentage of SAG production compared to the total output of East German industry was about 22 per cent. In several branches SAGs controlled larger portions of production, as shown in Table I. The figures indicate the particular interests of the Soviet Union, especially in heavy industries, as well as in the electrical industry, precision mechanics, and optics.

Table 1 Portion of SAG production in several branches of East Germany's industry 1947–1951[51]

Branch	1947	1948	1949	1950	1951
Energy	34,7	44,1	42,1	42,2	36,6
Mining	33,4	33,5	36,9	37,6	36,6
Metallurgical industry	58,5	44,5	36,2	31,7	26,7
Chemical industry	54,0	56,7	54,8	57,8	55,5
Engineering	39,4	28,0	29,6	26,4	27,8
Electrical industry	39,6	40,8	38,3	36,2	35,6
Precision mech./Optics	35,5	26,8	20,0	25,5	28,5
Building materials	15,3	15,8	14,5	12,9	12,0
Textiles	2,4	2,4	2,4	2,9	2,9
Cellulose/Paper[52]	.	6,4	6,1	5,8	5,4
Leichtindustrie[53]	—	0,2	2,4	1,9	1,9
Lumber industry	—	0,8	0,5	0,5	0,5
Food[54]	—	—	—	0,2	0,2
Printing industry	.	.	0,0	0,0	0,1
Industry, total[55]	**19,5**	**22,0**	**21,9**	**22,6**	**21,9**

Disagreement among the Allies about the SAGs proved advantageous to East Germany. Although in East Germany the dismantling of factories continued, the establishment of Soviet-owned companies prevented others from being dismantled and showed the willingness of the USSR to accept the retention of East Germany's most productive companies. The gross production per capita in SAG companies was 253.4 per cent higher than the average of industry in 1951.[56] Especially after the division of Germany following the introduction of two different currencies in 1948, East Germany's need to develop trade with the East grew rapidly.

On the other hand, the establishment of SAG companies based on the rules of the Civil Law shows Soviet flexibility, at least theoretically. Soviet actions in this regard were perhaps based on the following reasoning: either Germany could a) be organized as a capitalist country with capitalist-style Soviet-owned companies; or b) be divided into two Germanies with the step-by-step retransfer of those companies into East German state-owned property.

Indeed, the more East Germany was integrated into the Soviet bloc, the more SAG companies were retransferred into East German property. The first 74 SAGs were retransferred already in January 1947.[57] In April 1950 the Soviet Union announced its support for the entry of the GDR into Comecon (which actually happened on September 29, 1950) and was followed by the retransfer of

23 more SAGs.[58] In May 1952 the USSR agreed to sell another 66 SAGs to the GDR and announced the retransfer of the remaining companies (actually except the Wismut SAG) to take place in 1953.[59] This was followed by the Second SED Party Conference, where Walter Ulbricht announced the "construction of socialism in the GDR."

Finally, the break-up of traditional connections to Western markets made it necessary for East Germany to strengthen its ties to countries in the East. Although reparation orders influenced East Germany's general economic development according to the needs of the Soviet Union, those deliveries matched East Germany's interests as well. GDR's planning system granted, therefore, highest priority to the delivery of reparations. After 1954 reparation deliveries out of production were changed into "normal" export deliveries. Beside "normal" exports to the Soviet Union, which started in 1947 already and rose in the following years,[60] the reparations had created an interdependence between the needs of Soviet import agencies and the economy of East Germany. Between 1950 and 1955 over four-fifths of East Germany's exports went to Comecon countries. In 1954 alone, as compared to the figure from 1953, GDR's exports to these countries increased by about 31 per cent because reparations deliveries had come to an end.[61]

Allied policy toward western Germany also indicates a step-by-step change regarding its economic potential. The more West Germany became integrated into the Western alliance, the more it was possible to lift restrictions placed on its industry. Several reviews of the Level-of-Industry-Plan reduced this number of plants to be included as part of the war reparations. The first reduction of August 1947 decreased the number from about 1,500 to 859; the second reduction of April 1949 lowered this number still further to about 700.[62] In West Germany the entire dismantling program was, in effect, completed by the end of 1950.

Western Allied willingness to change its policy toward West Germany became much more obvious concerning the issue of production limitations and prohibitions. Whereas the three Western Allies reached an agreement in April 1949 whereby they recommended the removal of certain industries, such as those producing atomic materials, magnesium, beryllium, and war materials, only five months later they recommended simply the conversion of those industries to other types of production. The final step of this process resulted in the lifting in April 1951 of most of the limitations and prohibitions by the Agreement on Industrial Controls.[63]

The Western Allies undertook such step-by-step measures in order that western Germany could pay its own way. In the interests of limiting the amount of relief aid that western Germany required, its export potential would have to be increased. Although they placed restrictions on heavy industries, the Western Allies were even keen to see a gradual growth in West German contributions to West European defense.

Despite the remaining restrictions on some industries, the Western Allied policy inadvertently helped to lay the foundations for what came to be called the West German *Wirtschaftswunder* ("economic miracle"). By limiting the "dangerous" heavy industries, they indirectly promoted the development of "harmless" industries such as consumer goods. This had the effect that the economic recovery in West Germany was more obvious to its inhabitants than the economic recovery in East Germany was to its citizens.

Finally, to determine whether or not measures of the Allies were advantageous to Germany, we need consider carefully the case of German scientists and technicians who migrated to Allied countries. In contrast to Allied attitudes regarding material reparations, where increasingly confrontation determined the climate, there was rather a mixture of cooperation and competition in the case of German scientists. The allies managed rather quickly to reach agreement on controls of German scientific research and the exploitation of its results.[64] But it was not by chance that all Allies tried to hire or capture scientists from other zones.

The United States delayed its withdrawal from parts of the Soviet Zone until early summer 1945 and used its temporary authority in these areas to capture scientists and to remove documents, especially from Thuringia and Saxony-Anhalt.[65] The delayed withdrawal caused the Soviet authorities to lodge a protest,[66] and the capture of scientists caused the Russians to request the return of both scientists and documents.[67]

In a countermove, the Soviets in early 1946 invited scientists who had been moved from Thuringia to Heidenheim in Württemberg-Baden. Soviet authorities sent a flood of letters to "Heidenheim scientists." They also encouraged universities administrators to do the same. The major aim was to attract them back to East Germany.[68] The same happened to scientists living in the British Zone.

Within the community of German scientists captured in the American and British zones, there was a growing impression that a lot of scientists could get highly paid jobs in the Soviet Zone or in the USSR itself, while, as a German report of 1946 reads the "catastrophic development of the food situation in the British Zone" and the "so-called de-nazification" in the American Zone created a "general hopelessness," which weakened their "instinctive aversion to Bolshevism and Asiatentum."[69] After the escape of Zeiss scientist Herbert Kortum, a member of both the SS and NSDAP since 1931, from Heidenheim to Jena and his quick denazification by the Soviets, the American Military Government in Heidenheim/Württemberg rejected all requests from Kortum's wife to allow her and her family to return to Jena. These requests were turned down because "the return of this family might encourage other families to resort to the same method."[70]

Various kinds of material, i.e., letters, reports, and interviews, indicate that German scientists did not exclude a "Soviet option" when it came to employment possibilities in Allied countries.[71] High unemployment among the scientists who had ben moved to Heidenheim, made it more likely that Soviet or at least Soviet-directed offers from East Germany could be successful. Therefore, it is no wonder that American zonal military government ordered regional military governors to take measures to prevent the further hiring of scientists by the other Allies. The Office of Military Government in Württemberg-Baden, for example, was ordered in December 1946 to "contract every German scientist and technician requested by the War Department for shipment to the United States" and that "every necessary means must be used to induce him to sign the contract of his own free will."[72]

These efforts were aimed not a stabilizing the scientific community within Germany but at securing the largest portion of the "brain drain" to Allied countries. In the Soviet Zone the hunt for scientists aimed, above all, to strengthen the USSR's scientific community. First, in 1945 Soviet authorities "invited" only some German specialists to the Soviet Union. But during the night from October 21 to 22, 1946, Red Army units forced several hundred scientists and technicians, even those who had been lured away from West Germany, to migrate to the USSR.[73]

The latter event shows another difference between Soviet and Western Allied policies. In contrast to what happened in West Germany the Soviets forced larger groups of people to go to the Soviet Union. Lists of specialists can be reconstructed using lists of people who returned from the Soviet Union. They show that especially in the aircraft, precision mechanics, and optics industries, not only a few experts were captured but also engineers, *Meister*, and *Facharbeiter* (skilled workers).[74] Obviously, the Soviets intended to maintain working groups that had existed in Germany as teams in the USSR.

In preparation the events of October 1946 the Soviets had established what they called "Scientific Technical Bureaus" within or outside of the framework of Soviet Corporations in eastern Germany. Before forcing them to move to the Soviet Union the Soviets intended to assess the potential usefulness of scientists and technicians hired for the Scientific Technical Bureaus.

The evaluation of the scientists resulted in the decision of Soviet authorities either to bring the scientists to the USSR or to keep them in the employ of these "bureaus." Compared to the total number of employees working in these bureaus only a small portion was sent to the USSR in 1946. Those remaining in Germany continued to work on orders placed by SVAG authorities or the Soviet Academy of Sciences. This lasted until the transfer of the bureaus to the East Germans had been completed in 1954. Even after 1954, some of these new East German-owned research facilities filled Soviet orders, although this time based on regular contracts.[75]

On the other hand, among some German scientists there was interest in being hired rather than captured by one of the Allied powers. Working for either the Americans or the British was preferred. Under the conditions prevailing in Germany at that time — the destruction throughout Germany and severe restrictions on professional work — it made sense to them to resume prewar international contacts by accepting either offers or even orders to move to one of the victorious countries.

Moreover, the willingness of the Allies to form research groups, completely or at least partly comprised of German scientists, gave the latter a possibility to play one of the Allies off against the other. That means, for example, if a Nazi scientist failed to obtain his or her denazification certificate from one ally, he or she would try for it in another zone. West Berlin, which used to be called an "outpost of freedom," became the transfer point of "specialists" looking for the fastest route to denazification.[76] They had a chance to become "Cold War heroes," either in the West because they had "escaped communist dictatorship" or in the East because they had "escaped imperialist preparation for war."

CONCLUSION

The effects of the victors' political and economic treatment of defeated countries, following times of war, is a perennial topic of discussion. Whereas politicians and scholars of the victorious countries usually tend to underestimate the repercussions of these policies, those from the defeated countries often try to exaggerate the "damage" inflicted by them.

In the case of Germany after World War II, this tendency of interpreting the effects of Allied postwar policy in partisan terms is also very apparent. Politicians from Allied countries justified their immediate postwar policies toward Germany with constant reminders of the destruction caused by Germany during the war. In contrast, the Germans at this time experienced these very same policies as "plunder without end." There is no doubt that one could speak of a sort of "plunder" both with regard to the heavy dismantling of facilities as well as with regard to the confiscation of blueprints, patents, and trademarks.

The case of Germany also shows that the intentions of the victorious nations could quickly change. In the first place, the intended long-term prevention of its economic and scientific recovery lasted only few years after 1945. Compared to the consequences of World War I for Germany, the period of "uncertainty" concerning the negative effects of Allied policies toward the German economy was shorter.

Although dismantling continued until 1948 in the Soviet Zone and deliveries of reparations were sent until the end of 1953, the policies applied by all Allies toward their respective zones began already by 1947–48 to include steps that were guided by other purposes. Ultimately, in the long run the goal was to integrate the respective zones into either the Eastern or the Western postwar spheres of influence.

One reason for the quick change in the policy of all Allies toward Germany is obviously the fact that the victors of World War II had to deal with two Germanies instead of one, as was the case after World War I. This foiled the plans for Germany that the Allies had made during the war.

Another reason lies in the experiences the Allies had made in World War I. First, they feared that placing strict limitations on Germany's economic development would create revanchist ideas among the Germans. Second, there were fears that those limitations would curb potential Allied advantages to be gained by preserving German industrial capacity and integrating Germany into inchoate alliances.

As an irony of history, tainted scientists, i.e., those who had a Nazi past, could benefit from this policy. Many of them were reconnected with the international scientific community by offering their skills and talents to one of the Allied countries. In this way, Allied efforts to replace compromised teaching personnel or judges in the two German states for democratically or socialist-minded people was one side of the denazification coin; the enforced, or to put it better, the promoted brain drain of many Nazi experts was the other.

Whereas international law clearly defines the dismantling of machinery as a part of reparations, the opposite is true with regard to the confiscation of intellectual property and the migration of scientists. Intellectual property reveals its value only when it is used in the future. Since the use of dismantled machinery does not count as reparations, the same rule has to be applied to intellectual property, even though the removal of this kind of property damages the previous owners.

The migration of scientists and technicians also cannot be seen as "reparation services." Most of the specialists involved were looking forward to opportunities to emigrate from Germany at least for a couple of years. Even those scientists who moved to the Soviet Union exchanged harsh conditions in Germany with a kind of "golden cage" in which they had to live.

When the first scientists began returning from the USSR, they created a major problem for the East German regime. Many of the scientists who returned for good demanded that the country provide working conditions, which included special salaries, similar to those they had enjoyed during their stay in the Soviet Union. For that purpose, East Germany created the system of *Einzelverträge* (i.e. special contracts). These included various measures to privilege returnees from the Soviet Union. Some of the scientists even got their own institutes, organized according to their wishes.[77]

At the same time, the *Einzelverträge* were intended to contribute to the efforts of East German authorities to prevent scientists from moving to West Germany, which had already intensified efforts to promote the remigration of scientists who had gone to Allied countries, especially to the Soviet Union.

It is more likely than not that migration to Allied countries even aided the return of Germany to the international scientific community. The case of West Germany demonstrates that reestablishment of prewar international scientific contacts was possible because it was in interest of the Western Allies to maintain and reintegrate West German industrial and scientific potential. The same was not true, however, in the case of East Germany's scientific community. Although officials of the Soviet Control Commission, established in connection with the foundation of the German Democratic Republic in late 1949, directly and indirectly influenced East Germany's research policy,[78] the remaining distrust toward German scientists prevented the establishment of a "socialist-orientated" Eastern international scientific community.

NOTES

1 Horst Barthel, *Die wirtschaftlichen Ausgangsbedingungen der DDR. Zur Wirtschaftsentwicklung auf dem Gebiet der DDR 1945–1949/50* (Berlin: Akademie-Verlag, 1979).

2 John Gimbel, *Science, Technology, and Reparations, Exploitation and Plunder in Postwar Germany* (Stanford, Calif.: Stanford University Press, 1990).

3 Ulrich Albrecht/Andreas Heinemann-Grüder/Arend Wellmann, Die Spezialisten, *Deutsche Naturwissenschaftler und Techniker in der Sowjetunion nach 1945* (Berlin: Dietz 1992). Thereafter: Albrecht, "Spezialisten."

4 Tom Bower, *The Paperclip Conspiracy, The Hunt for the Nazi-Scientists* (Boston/Toronto: Little, Brown, 1987).

5 Linda Hunt, *Secret Agenda, The United States Government, Nazi Scientists, and Project Paperclip, 1945 to 1990* (New York: St. Martin's Press, 1991).

6 Josef Foschepoth, "Konflikte in der Reparationspolitik der Alliierten," in: Josef Foschepoth (ed.), *Kalter King und Duetsche Frage, Deutschland im Widerstreit der Mächte* (Göttingen/Zürich: Vandenhoeck & Ruprecht, 1985), 175–197 on pages 177–179.

7 *History of Field Information Agency, Technical (FIAT). Period 8 May 1945–30 June 1946* (thereafter: "History of FIAT"), in: National Archives and Record Administration, Washington, D.C. (thereafter: NARA) Record Group 407, Box 414.

8 See: Yevgeny Varga: "Wosmescheniye uscherba gitlerowskoy Germaniyei i yeyo soobschnikami" (The Reparation of Damage of Hitlerite Germany and her Accomplices), in: *Voyna i rabotschi klass* (War and the Working Class), 1943, *10*, 4–10 on page 10.

9 *Interview of Matthias Judt* with Ivan Beydin (after 1945 responsible for youth organizations within the Soviet Military Administration in Germany — SVAG), August 31, 1992, Gosen near Berlin; see also: Michail Smirjagha: "Die Rote Armee in Deutschland in Jahre 1945," in: Peter Jahn, Reinhard Rürup (eds.), *Erobern und Vernichten. Der Krieg gegen die Sowjetunion 1941–1945*, (Berlin: Argon, 1991) 200–210 on page 202.

10 Stiftung Archiv der Parteien und Massenorganisationen der DDR in Bundesarchiv, Zentrales Parteiarchiv der SED (thereafter: StAPM-BA, SED), NL 36 (Wilhelm Pieck papers), No. 734, passim.

11 Proposals of Henry Morgenthau for the treatment of Germany after capitulation, September 1944, quoted in: Gustav Harmssen, *Am Abend der Demontage. Sechs Jahre Reparationspolitik* (Bremen: Friedrich Trüjen Verlag, 1951), 161–162.

12 See: "Memorandum for the President" by the Secretary of State, Riddleberger, March 29, 1944, in *Foreign Relations of the United States, Diplomatic Papers. The Conferences at Malta and Yalta* 1945 (*thereafter*:

FRUS 1945), (Washington, D.C.: United States Government Printing Office, 1955), 157–158.

13 Directive to Commander-in-Chief of United States Forces of Occupation Regarding the Military Government of Germany (JCS 1067), quoted in: James Warburg, *Germany. Bridge or Battleground* (London-Toronto: William Heinemann, 1947), 283. *Thereafter*: Warburg, "Germany."

14 JCS 1067, in: Warburg, "Germany," 289.

15 Paper by B.K. Blount "The Future of German Science," 1945, NARA, RG 260, Office of Military Government, U.S. (*thereafter*: OMGUS), Office of Military Government Wuerttemberg-Baden (*thereafter*: OMG W-B), LSO, Box 179.

16 Willi A. Boelke, *Die Kosten von Hitlers Krieg, Kriegsfinanzierung und finanzielles Kriegserbe in Deutschland 1933–1948*, (Paderborn: Schöningh, 1985), 141.

17 Recalculated according to the figures, calculated by Thomas Kuczynski: "Allgemeine Bedingungen und Tendenzen der Entwicklung," in: *Produktivkräfte in Deutschland 1917/18 bis 1945* (Berlin: Akademie-Verlag, 1988), 15.

18 Calculated according to *Statistisches Jahrbuch für das Deutsche Reich 1941/42*, 665, quoted in Rolf Wagenführ, *Die deutsche Industrie im Kriege 1939–1945* (*thereafter*: Wagenführ, "Industrie") (Berlin: Duncker & Humblot, 1954), 20.

19 Calculated according to database of "Zusatzerhebung der Reichsgruppe Industrie zum Industriebericht," quoted in Wagenführ, "Industrie," 158.

20 "Comminique Issued at the End of the Conference. Report of the Crimea Conference," in: FRUS 1945, 970.

21 See: Letter of Clay to Joint Chiefs of Staff, June 6, 1945, edited in: Jean Howard Smith, *The Papers of General Lucius D. Clay, Germany 1945–1949, Volume One* (*thereafter*: Smith, "Clay Papers") (Bloomington/London: Indiana University Press, 1974), 18; Albrecht, "Spezialisten," 27; *Amtsblatt des Kontrollrats in Deutschland*, No. 1, Berlin, October 29, 1945, 8–10.

22 Quoted and calculated (unemployment rate) according to "Expertenbericht zum Plan der Reparationen und des Nachkriegsstandes der deutschen Wirtschaft (Industrieplan vom 29. März 1946)" of the *Statistisches Zentralamt der Sowjetischen Besatzungszone*, 1947, in: StAMP-BA, SED, IV 2/6.02 (Central Committee's *Abteilung Wirtschaftspolitik*), No. 45, 39 and 51.

23 See: StAMP-BA, SED, IV 2/6.02/52, passim. *The Wirtschaftskammer Dresden/Sachsen* sent 12 reports until August 31, 1945, and another 5 until October 24, 1945.

24 See: Bundesarchiv, Abteilungen Potsdam (quoted thereafter as BArchP), DO-3 (*Zentrale Deutsche Kommission für Sequestrierung und Beschlagnahme*), Nos. 7 and 1395–1408, passim.

25 Matthias Judt, "Die Demontagen und die Bildung von Sowjetischen Aktiengesellschaften als Potentialverluste", in: Elke Scherstjanoi (ed.), Provisorium für längstens ein Jahr. Protokoll des Kolloquiums "Die Grüdung der DDR", (Berlin: Akademie-Verlag, 1993), 220.

26 Allied Control Authority (*thereafter*: ACA), Control Council, Law No. 25, in: NARA, RG 260, OMGUS, ACA, CONL, "Master Files," Box 104.

27 ACA, Control Council, Law No. 29, 1, in: NARA, RG 260, OMGUS, ACA, CONL, "Master Files," Box 104.

28 "History of FIAT," 12.

29 See: "Operational Procedures Concerning Dismantling, Packing, and Transportation of Plants Subject to Deliveries on account of Reparations," in: NARA RG 260, OMGUS, ACA, Reparations, Deliveries and Restitution Directorate (*thereafter*: DRDR), Box 365.

30 "Minutes of the Sixteenth Meeting (of the DRDR-M.J.), 6 December 1945," in: NARA RG 260, OMGUS, ACA, DRDR, Box 365.

31 All figures in 1938 U.S. dollars.

32 See: Jörg Fisch, *Reparationen nach dem Zweiten Weltkrieg* (München, C.H. Beck, 1992), 217. *Thereafter*: Fisch, "Reparationen."

33 Calculated according to: Fisch, "Reparationen," 222, under exclusion of the occupational costs in 1953 (already counted in the figure for the period between 1945 and 1953).

34 According to international law, legal confiscations are limited to war equipment according to the international law. In addition to that the Soviet Union declared other confiscations (the so-called "trophy goods") to spoils of war. This included raw material, blue prints and others.

35 In April 1946, the Soviets established the *Deutsch-Russische Transport AG* (Derutra). It was the only company that was allowed to transport these goods. This Soviet-owned company only sometimes gave orders to other transportation businesses.

36 The Soviet Union ordered reparations based on German prices as of 1944, which were called *Meßwerte* in the reports of GDR's reparation agency. For East Germany the Meßwerte expressed the 1944 wholesale prices (see: (confidential) *Mitteilungen in Reparationsangelegenheiten Nr. 2/50*, in: StAMPO-BA, SED, NL 90 (Otto Grotewohl papers), No. 338, 156), which were 13.6 per cent higher than in 1938 (see: "Index der Großhandelspreise in der DDR (1936–100)," in: BArchP, DE-1 (*Staatliche Plankommission*), Nr. 28332, 13). Unless otherwise indicated, the figures were recalculated into prices as of 1938. The further calculation uses the exchange rate $/Reichsmark 1:2.5.

37 According to the *Reparationskartei/Ost* (Card Index of Reparations in East Germany) of the (West German) Economics Ministry, East Germany lost facilities and material with the value of 6.09 billion Reichsmarks (1938). For estimates see: (unpublished expertise) Doris Cornelsen/Maria Elisabeth Ruban/Maria Elisabeth Ruban/Dieter Teichmann, *Kreigsschäden und Nachkriegsbelastung in der Bundesrepublik Deutschland und in der*

DDR (Berlin: Deutsches Institut für Wirtschaftsforschung, thereafter: DIW, 1972) 116a.

38 The total amount expressed in German wholesale prices as of 1944 is 7.4242 billion Reichsmarks. This calculation bases on estimates, available for the years 1945–1947, and reports of East Germany's *Amt für Reparationen*, indicating the figures for 1948–1953. Sources: 1945–1947: Lothar Baar/Rainer Karlsch/Werner Matschke, Kreigsfolgen und *Kriegslasten Deutschlands. Zerstörungen, Demontagen und Reparationen* (Berlin: Humboldt-University, 1993), 66 (*thereafter*: Baar, "Kriegsfolgen"); 1948: "Erfüllung der Anordnungsbestellungen auf Rechnung Reparationen nach dem Stand vom 31. Dezember 1948," in: Bundesarchiv, Abteilungen Potsdam, Außenstelle Coswig/Anhalt (*thereafter*: BArchC), DC-2 (*Amt für Reparationen*), 1; 1949: "Bericht über die Erfüllung der Reparationsverpflichtungen im Jahre 1949," in: BArchC, DC-2, No. 17088, 3; 1950: "Bericht über die Erfüllung der Reparationsverpflichtungen in Jahre 1950," in: BArchC, DC-2, No. 2987, 36; 1951: calculated according to the percentage of fulfillment of the 1951 reparation plan (see: "Bericht über die Erfüllung der Reparationsverpflichtungen im Jahre 1951," in: BArchC, DC-2, No. 17098, 1 and 7); 1952: calculated according to the percentage of fulfillment of the 1952 reparation plan (see: "Bericht über die Erfüllung der Reparationsverpflichtungen im Jahre 1952," in: BArchC, DC-2, No. 17103, 36); 1953: "Bericht über die Erfüllung der Reparationsverpflichtungen im Jahre 1953 (R1-Bericht)," in: BArchC, DC-2, No. 17410, 2.

39 Calculated according to the estimate in Baar, "Kriegsfolgen," 76, that the Wismut' mining cost 4.5 billion Reichsmarks/Marks in prices as of 1944.

40 Order lists for deliveries to Soviet troops in East Germany show the use of the Meßwerte, too. Furthermore, Fisch quotes Rupp, who pointed out, that East Germany spent 9 billion Marks in costs of occupation until 1950. Fisch quotes a speech of East Germany's Prime Minister, Otto Grotewohl, in the country's parliament, the *Volkskammer*, on August 25, 1953, too, where Grotewohl gave the figure of 10.025 million Marks for these costs for the period between 1949 and 1953. Sources: Fisch, "Reparationen," 194; Franz Rupp, *Die Reparationsleistungen der sowjetischen Besatzungszone*, (Bonn: Deutscher Bundesverlag, 1951), 19; "Die Entwicklung der Volkswirtschaft und Kultur in der Deutschen Demokratischen Republik in den Jahren 1950 bis 1953," in: BArchP, DE-1, No. 422, 39.

41 The calculation follows the figure, used in Baar, "Kriegsfolgen," 72, that the repurchased SAG companies had a value of 1.085 billion Reichsmarks in prices as of 1944.

42 In Baar, "Kreigsfolgen," 36, it is estimated that goods with the value of 1 billion Reichsmarks in prices as of 1944 were confiscated, illegally.

43 There are only estimates about *Besatzungsgeld* available, but information, too that the Soviet Union did not use the whole amount of money for the purchase of goods in East Germany. In Baar, "Kriegsfolgen," 80, informa-

tion is given that Soviet authorities burned greater amounts of Besatzungsgeld and transferred 2 billion Reichsmarks to the German Economical Commission right before the East German currency reform.

44 It is estimated that 3.1 billion Reichsmarks circulated as *Besatzungsgeld.*

45 The calculation follows the estimate in Baar, "Kriegsfolgen," 84, that East Germany lost 1 billion Reichsmarks (as of 1938) in its foreign trade between 1945 and 1956.

46 Rainer Karlsch estimates that the Derutra received 500 billion Marks out of zonal and *Länder* budgets (see: Rainer Karlsch, *Allein bezahlt? Die Reparationsleistungen der SBZ/DDR 1945–1953* (Berlin: Ch. Links Verlag, 1993), 216). Since there is no further information on the applied price basis, we assume current prices, to be careful. The further calculation follows the estimate that between 1945 and 1953 the average wholesale prices were 50 per cent higher than the prices as of 1938 (see: Baar, "Kriegsfolgen,"[68]).

47 The calculation follows the estimates and calculation methods in Baar, "Kriegsfolgen," 67–68 indicating that Soviet Trading Companies in eastern Germany made hidden purchases with a value of about 1 billion Marks in current prices.

48 Werner Abelshauser, "Probleme des Wiederaufbaus der westdeutschen Wirtschaft," in: Heinrich August Winkler (ed.), *Politische Weichenstellungen im Nachkriegsdeutschland 1945–1953*, (Göttingen: Vandenhoeck & Ruprecht, 1979), 208–253 on page 241.

49 See: "Gesellschaftliches Gesamtprodukt nach Wirtschaftszweigen und Wirtschaftsbereichen 1936, 1950–1957, in: BArchC, DE-1-St (Staatliche Zentralverwaltung für Statistik), No. 0643, 2.

50 See: Rolf Krengel, *Anlagevermögen, Produktion und Beschäftigung im Gebiet der Bundesrepublik von 1924 bis 1956* (= DIW Sonderhefte NF No. 42) (Berlin: DIW, 1958), 94–95.

51 1947–1950: quoted according to Werner Krause, *Die Entstehung des Volkseigentums in der Industrie der DDR* (Berlin; Die Wirtschaft, 1958) (*thereafter*: Krause, "Volkseigentum") 104. 1951: calculated according to "Volkseinkommen, Mappe XIV," in: BArchP, DE-1, Nr. 28332, 21–26 and 28–29.

52 No figure available for 1947.

53 Term used in East German statistical data, meaning clothing, leather, shoe and fur industries, combined.

54 1951: without production of *Handelsorganisation* (state-owned retailing company with production facilities).

55 1951: without production of *Handelsorganisation* in food industries.

56 Calculated according to the database of "Bruttoproduktionswert je Beschäftigten, der in der Produktion Beschäftigten und der Produktionsarbeiter in der Industrie der DDR," in: "Volkseinkommen, Mappe XIV," BArchP, DE-1, No. 28332, 170.

57 See: Krause, "Volkseigentum," 101.
58 For the GDR's entry into the Comecon see: "Bemerkungen Semjonows am 7.4.1950, in: StAMP-BA, SED, NL 36, No. 736, 106. For the re-transfer of SAGs see: BArchP, DO-3, No. 1428, 59 and 62–63.
59 See: StAPM-BA, SED, NL 36, No. 736, 310–311.
60 See: "Entwicklung des Außenhandels der Deutschen Demokratischen Republik mit UdSSR," in: BArchC, DE-1-St, No. 5861, 15.
61 Calculated according to Hellmuth Kalus, *Wirtschaftszahlen aus der SBZ*, (Bonn: Deutscher Bundesverlag, 1958), 83 (based on official statistical data of the GDR).
62 Office of the U.S. High Commissioner for Germany, *Report on Germany, September 21, 1949–July 31, 1952*, (Cologne: printed by Greven & Bechthold, 1952), 123.
63 *ibid.*, 124.
64 See: paper "Technical and Scientific Research in Germany after the War" of June (?) 1945, which recommended the establishment of a Central Intelligence Objectives Bureau in Berlin. In: NARA, RG 260, OMGUS, FIAT, Administration Records, Box 7.
65 See: Arnim Hermann, *Nur der Name war geblieben. Die abenteuerliche Geschichte der Firma Carl Zeiss*, (Stuttgart: Deutsche Verlagsanstalt, 1989), 12; Paul G. Esche/Herbert E. Kessler, *Der VEB Carl Zeiss Jena. Seine äussere Entwicklung von der Gründung bis zur Gegenwart* (= *Schriften des Stadtmuseums Jena, Band 3*), (Weimar: Keipert, 1966), 61.
66 Still at the meeting of the Zone Commanders in Berlin Karlshorst (at the Headquarters of Zhukov) no agreement on the withdrawal was reached, although "Zhukov apparently wanted an answer rather than a discussion" on this topic (see: letter from Clay for Joint Chiefs of Staff, June 6, 1945, in: Smith, "Clay Papers," 18–20).
67 Letter of Sokolovsky to Clay, October 19, 1945; answer: Letter of Clay to Sokolovsky, December 19, 1945; Letter of Shabalin (chief of Economics Division of SVAG) to Draper, October 23, 1945; answers: Letter of Draper to Shabalin, November 10, 1945 (with information, that Director of Intelligence of OMGUS would act), Letter of Bryan Conrad (Director of Intelligence) to Shabalin, December 7, 1945, all in: NARA, RG 260, Chief of Staffs, FIAT Administration Records, Box 7, passim.
68 See for example: Letter of Struck, dean of the Faculty of Physics and Mathematics of Jena University, to colleagues at Heidenheim, June 6, 1946; Extract of a letter from Ernst Sattler (Oranienburg/Soviet Zone) to Dr. Hellmut Sattler (Niederstotzingen/American Zone), July 6, 1946, in: NARA, RG 260, OMGUS, OMG W-B, LSD, Box 179.
69 See unsigned paper "Die unterschiedlichen Berufsaussichten der Forscher und Techniker in den Besatzungszonen Deutschlands und ihre Folgen," March 26, 1946 (focusing on the British Zone), 4–5, in: NARA, RG 260, OMGUS, OMG W-B, LOS, Box 179.

70 Letter of Robert N. Thrap, L & S Officer of OMG Heidenheim to Headquarters USFET, August 7, 1946, in: NARA, RG 260, OMGUS, OMG Wuerttemberg-Baden, Field Relations Division, LSO Heidenheim, Records Relating to German Scientists (RG 260, OMGUS, OMG W-B, LSO), Box 180.

71 NARA, RG 260, OMGUS, OMG W-B, LSO, Box 179; *Interview of Matthias Judt with Werner Holzmüller* (formerly physicist at Leipzig University), March 27, 1992 in Böhlitz-Ehrenberg near Leipzig; Interview of Hermann Vinke with Werner Albring "Wir nannten uns Friedensgefangene," in: Hermann Vinke (ed.): Werner Albring, *Gorodomlia. Deutsche Raketenforscher in Rußland*, (Hamburg/Zürich: Luchterhand Literaturverlag, 1991), 15.

72 OMGUS to Director OMG Wuerttemberg-Baden, December 9, 1946, in: NARA, RG 260, OMGUS, OMG W-B, LSO, Box 179.

73 "Interview with Werner Holzmüller"; Mitteilung des Geschäftführers der Industrie- und Handelskammer Thüringen, Siewerth, October 22, 1946, in: StAMP-BA, SED, VI 2/6.02/52, 52 and 330; *Sozialdemokrat* No. 93, October 25, 1946.

74 StAMP-BA, SED 2/13 (Central Committee's *Abteilung Staat und Recht*), No. 389, various reports on the remigration of Specialists, passim.

75 "Themen, die im Auftrage der UdSSR durchgeführt werden" (1954 and 1955), In: BArchC, DF-4 (Ministerium für Wissenschaft und Technik der DDR), No. 720, 7.

76 See for example: "Industrial Situation Report No. IV in the US Sector of Berlin as of 31 December 1947," in: NARA, RG 260, OMGUS, OMG Berlin Sector, Office of Economic Affairs, Box 583, in which the Chief of the Industry Section of OMG Berlin Economics Branch, Swanson, reported, that an "increasing number of first-class technicians were lost [by the US sector -M.J.] to the USSR and other countries in its sphere of influence. They were attracted by promises of better food rations or, in the case of former Nazis, by promises of exemption from discrimination."

77 Examples are Peter-Adolph Thiessen and Manfred von Ardenne. The latter started the preparations for his own institute already, when he still was in the USSR. See: Letter of Lange to the Mayor of Dresden, August 18, 1952, "Betrifft: Freistellung des Hauses Dresden-Weißer Hirsch, Plattleite 27–29," in: BArchC, DF-4, No. 464.

78 The research plans had to be "defended" before the Soviet Control Commission, which could demand changes with regard to research topics, the funding and other things. See for example "Niederschrift zur Besprechung mit Herrn Postnikow am 14.4.51 über den Forschungs — und Entwicklungsplan 1951" (of May 5, 1951), in: BArchC, DF-4, No. 118, 593–55.

CHAPTER 4

The Nazification and Denazification of Physics

Mark Walker[1]

The effect of political ideology on science and technology remains an important question for historians. The Third Reich is a prime example of a dangerous ideological environment hostile to science. Unfortunately, most historical work on German science and technology either stops or begins in the spring of 1945, thereby denying or at least obscuring the legacy of National Socialism for postwar science.[2] This article will examine both the nazification and the denazification of physics in Germany. Although the fate of physics was not identical to that of the other sciences or engineering, this case study can nevertheless serve to illuminate science and technology under Hitler.

THE MYTH OF "NAZI PHYSICS"

The concept of myth is very important for understanding nazification and denazification in general, but perhaps especially for physics. There are two related obstacles to our understanding. First of all, there is a persistent myth associated with physics under National Socialism, what one might call the myth of "Nazi Physics": the claim that the interaction between National Socialism and German physicists can and should be both equated with, and limited to the so-called "*Deutsche Physik*" movement.[3] However, merely pointing to the existence of this myth is not enough. Why this legend was created, propagated, and — most importantly — believed, should be investigated as well.

The second obstacle transcends physics or even science under Hitler and represents a fundamental problem for an understanding of National Socialism and its legacy in general: the persistent and virulent use of the Janus-like combination of hagiography or demonization, the black-and-white characterization of scientists — like other professions and social groups — as fitting into three mutually exclusive categories, "Nazi," "anti-Nazi," or neither one nor the other. One could also label these categories "Heaven," "Purgatory," and "Hell," for they are based on the timeless, if sometimes simplistic theme of the struggle between good and evil.

There are at least three reasons why this Janus-like characterization of scientists persists. First of all, the historiography has been decisively influenced by practising or retired scientists, who sometimes lack the necessary critical distance when investigating the less noble aspects of their profession. Secondly, many historians insist upon the validity and even necessity of designating who were the "Nazis" and who were the "anti-Nazis," even though such designations may lead to black-and-white interpretations of the complexity of life under National Socialism. Finally, some historians and scientists are consciously or unconsciously hypocritical: they condemn hagiography or demonization by others, but when it comes to their own heroes or villains, often the subjects of years of their own work and reflection, they make exceptions and fail to submit them to the same critical analysis reserved for others.

Rather than trying to place individuals, or for that matter, institutions into such exclusive categories, one may use the model of a spectrum of attitudes and actions, of shades of gray. Here each case lies somewhere on the spectrum, influenced both by the light and the dark. Although the two ends of the spectrum can be thought of as "Nazi" and "anti-Nazi," in practice these extremes are almost never reached.

NAZIFICATION AND MILITARIZATION

If German science was denazified, then it must first of all have been nazified, which may be defined as follows: the effective, significant, and conscious collaboration by scientists with portions of National Socialist policy. Here nazification is best applied to large groups, such as the discipline of physics, and would be problematic if used for individuals. Given the great amount of variation among physicists and scientists during the Third Reich, of course there was significant variation as well with regard to the form and course of their interactions with National Socialism.

Both the purge of the German civil service and of German science are well known, but historians of science have sometimes failed to recognize that the purge of scientists, at least for academics, was an *automatic* result of the greater civil service purge and was *not* a conscious National Socialist policy against science in particular. The National Socialists were too disdainful of science to place it high on their list of things to "synchronize." Education in general and university education in particular were priorities for Germany's new leaders, but in this regard scientists were treated no differently from their non-scientific colleagues.

Albert Einstein here represents the exception that proves the rule.[4] He was a respected scientist before World War I, but the unusual popularization of his theory of relativity during and after the war, combined with his unconventional personal political stance as pacifist and internationalist, made him into a cultural icon and political symbol during the Weimar Republic. The National Socialists singled Einstein out for their wrath precisely because his political significance transcended his science.

The best known example of the nazification of physics is of course the *Deutsche Physik* movement. During the Weimar Republic, two prominent Nobel laureates, Philipp Lenard and Johannes Stark, became alienated both professionally and politically and channeled their frustration, like many Weimar conservatives, into support for Adolf Hitler's fledgling National Socialist German Workers Party. Lenard's and Stark's struggle against the influence of modern physics was at first restricted to professional channels, but by 1933 both scientists publicly used the virulent anti-Semitic rhetoric so common on the political right and so integral a part of National Socialism. Perhaps most important, the public support provided by Lenard and Stark to the National Socialist movement early in the Weimar republic was rewarded with political influence at the beginning of the Third Reich, however, influence which they soon squandered.

It is important to recognize how the National Socialist "synchronization" functioned, why it was successful, and what pattern it followed. First of all, Hitler and his followers needed and received assistance from influential members of Germany's conservative elites — including scientists. Secondly, and most important for this subject, the policy of "synchronization" was neither centrally planned, nor centrally coordinated, nor centrally implemented. Instead the "seizure of power" was characterized by uncoordinated and often unsolicited pressure from the National Socialist rank-and-file party members and Storm Troopers (SA). This often violent pressure was then exploited by the National Socialist authorities to eliminate step-by-step all opposition.

But since the National Socialist leadership also wished to present an image of a peaceful, orderly society under their control, such "revolution from below" eventually became counterproductive. Hitler publicly called for "evolution, not revolution," a thinly-veiled threat to his own followers. When in 1934 the SA leadership persisted in its calls for a "second revolution" which would have benefited in particular the lower levels of the National Socialist movement, Ernst Röhm and the majority of the SA leadership were eliminated in a bloody "night of the long knives" and their calls for change silenced.

The nazification of German physics followed this "SA model" and its four stages: (1) "revolution from below"; (2) "evolution, not revolution"; (3) "second revolution"; and (4) finally the National Socialist "revolution" devouring its own "children." With some modification, this model can be applied to the other sciences as well as to engineering, although the timing for the various stages varied and the results for scientists and engineers were rarely murderous.[5]

The *Deutsche Physik* movement in general and Johannes Stark in particular tried to take over the German physics community by means of a very effective campaign of character assassination. During the first few years of the Third

Reich Stark became the president of the Germany Research Foundation and the Imperial Physical-Technical Institute, respectively the source of most governmental funding for science and one of the best-equipped research centers for physics, and secured a promise that he would be consulted before physics professorships were filled.

However, by 1936 Stark and his allies were beginning to get in the way of other, more influential forces within the National Socialist movement. Stark in particular now found his efforts to achieve a "second revolution" in German physics, which would go beyond the initial purge of the civil service, were being opposed by officials in the Ministry of Education and SS. In 1936 Stark was forced to resign from the German Research Foundation and barely managed to fend off efforts to throw him out of the party. In 1939 he retired from the Imperial Physical-Technical Institute. Stark saw his influence steadily fade when the war began and as it progressed. Indeed, Stark became so alienated that he took the very unusual and potentially dangerous step of trying to resign from the National Socialist party during the last years of the war.[6]

Stark and *Deutsche Physik* were first opposed and then neutralized by other and stronger parts of the National Socialist movement because the long-standing goals of the former conflicted with the new ambitions of the latter. It was the established physics community, and not the scientifically sterile *Deutsche Physik*, which could and did effectively contribute to the National Socialist goals of rearmament, the creation of a racially "pure" society, cultural imperialism, geographic expansion through military aggression, and finally genocide.[7]

Perhaps the best known victory by the *Deutsche Physik* movement, the so-called "Sommerfeld succession," also marked the high point of its influence.[8] Stark used articles in the National Socialist newspaper *Völkischer Beobachter* and SS weekly *Das Schwarze Korps* to attack Heisenberg as a "white Jew" and "Jewish in spirit" in order to prevent him from becoming Arnold Sommerfeld's successor as Professor of Theoretical Physics at the University of Munich. Heisenberg was in fact denied this appointment, but this was a good example of *Deutsche Physik* winning the battle, but thereby losing the war.

The attacks on Heisenberg and the modern physics he symbolized set in motion both a political investigation by the SS of Heisenberg's character and a series of successful efforts by leading applied and industrial scientists to convince industrialists and leaders of the armed forces that the political attacks by *Deutsche Physik* were endangering the quality of physics education and thereby of Germany's ability to train the technicians, engineers, and scientists it needed. The SS also decided to back Heisenberg because of his potential to train another generation of physicists. However, there was a price: Heisenberg had henceforth to distinguish between support for a scientific theory and for a scientist. The theory of relativity and quantum mechanics could be used and taught, but Einstein was to be attacked or at least ignored.

Perhaps more important, the decision by Heisenberg and his colleagues to work within the system also meant that they inevitably collaborated with it. The

National Socialist state gave them something they wanted; they would have to return the favor. Ironically it was not the *Deutsche Physik* movement itself that nazified physics, rather the successful struggle by the established physics community against Lenard and Stark and the consequential collaboration in entailed. Similarly, war, rearmament, and the opportunities they offered scientists nazified industrial physics in Germany.

However, the question remains: *why* was German physics nazified *in this way*? The adherents of *Deutsche Physik* simply tried to expand their influence within the German physics community any way they could, and initially their strategy appeared successful. The established German physics community could easily find influential and sympathetic patrons within the sometimes chaotic and contradictory polycratic structure of the Third Reich. This support was sometimes given for reasons of principle, sometimes as a cynical, tactical stance within the shifting politics of the National Socialist state, but no matter why these patrons chose to side against *Deutsche Physik*, some of them were in a very strong position to do so. Finally, the overwhelming majority of German physicists allied themselves with, or submitted themselves to forces within the Third Reich because, when compared to the extreme politicization represented by *Deutsche Physik*, these scientists saw this course as the *lesser evil*.

DENAZIFICATION AND DEMILITARIZATION

The Allies pledged themselves to denazify and demilitarize Germany once the Third Reich had been defeated. Unfortunately, Germany's new rulers agreed only on their *intention* to purge public life entirely of Nazi influence.[9] There were grave differences from the very beginning with regard to the timing and scope of denazification in practice, differences based in turn on fundamentally different analyses of the causes and supporters of Nazism. In the Soviet zone denazification played an important role in the construction of a new social order based on the Soviet model. In the western zones denazification was essentially restricted to a comprehensive political purge of personnel, for which — in contrast to the Soviet zone — the economic sphere basically remained untouched. Denazification was a pillar of American occupation policy, but was much less important for the British and the French.

Initially, the occupying forces ran the denazification themselves, often with catastrophic consequences for public administration and the economy. Mere membership in the National Socialist German Workers Party or an ancillary organization could be grounds for dismissal pending the outcome of denazification, a policy which had the predictable effect of creating solidarity among those individuals facing this blanket threat. However, denazification was quickly turned over to the Germans themselves to run, both in order to save money and because only Germans were in a position to make a differentiated judgment of conduct under National Socialism. Denazification was now recast as a judgment of personal responsibility, not mere membership in a political organization, and led to what Lutz Nietzhammer has called the "Factory for Fellow Travelers" (*Mitläuferfabrik*).[10] The four powers decided to wrap up denazification by early 1948, at a time when, in the West at least, the "little Nazis" had been through the process, but the "bigger fish" remained relatively unscathed.

Clemens Vollnhals has written with regard to denazification that "without the agreement and cooperation of the Germans, a political purge could be administratively ordered, but not effectively and lastingly carried out."[11] Of course, this judgment also holds for the purge of German society by the National Socialists in 1933. Perhaps the fundamental question was the *meaning* of denazification: was it intended to neutralize the threat of a Nazi revival, or to punish previous Nazi conduct?

In retrospect, denazification seems to have been doomed to failure. Geoffrey Giles has rightly pointed out that only the negative, punishing side of denazification was effective: purging the old was an easy task, but filling the void created was an altogether different and infinitely more difficult one.[12] Since the western zones were at least on their way to becoming democracies, their politicians had to cater to the majority will, which was hardly self-critical or enthusiastic about denazification. The shortage of democracy in the education system, including the universities, made it harder to do without the existing anti-democratic elements. Thus the occupation officials were faced with a dilemma: should democracy be introduced into the universities, even though they suspected that a majority of academic were anti-democratic?

In any case, scientists and in particular physicists were nothing special with respect to denazification, rather were subject to the general denazification of the universities if they wanted an academic career. The denazification of 1945 to 1948 was as much of a political purge as was the nazification that had begun in 1933. In 1945 no one asked whether these scientists were good physicists or qualified teachers; they were judged by *political* criteria. Far fewer physicists were purged after 1945 than 1933, due in part to the fact that the dismissals and expulsions of 1933 were sometimes *racial* in nature, a criterion not employed after 1945. But this factor does not suffice to explain the stark contrast between 1933 and 1945. There was a severe shortage of physicists in Germany after World War II, so that pragmatism is another part of the explanation.

But here again myth can help illuminate this process. According to the usual postwar "party line" of the established German physics community, Germany physics had fallen behind American science because "the Nazis ruined German science." But the same physicists often insisted that most of the physicists who were in place after 1949 were of very high quality. If the Nazis ruined physics, then one would expect that many of the physics professors, who began their careers after 1933, would be incompetents who owed their appointment only to political influence; conversely, if the postwar academic physics community was of high quality, then one could hardly say that the Nazis had ruined German science.

When attrition due to aging and the employment of physicists by the victorious powers are taken into account, the very small group of physicists purged from academia after 1945 is practically equivalent to the former adherents of *Deutsche Physik*. It is no surprise that denazification barely touched physics — it barely touched almost everything — but that in no way explains why only *Deutsche Physik* was purged. No other subset of German physicists, including SS physicists, was punished so thoroughly or consequently.

The denazification of physics can be illuminated by the same model used for its nazification, only run backwards: (1) the established physics community collaborates with forces within the occupying powers in order to punish the former supporters of *Deutsche Physik*; (2) *Deutsche Physik* resists, ironically arguing accurately that it could not be equated with the total and worst excesses of National Socialism; (3) the occupying powers accept the established physics community "party line" and single out *Deutsche Physik* for punishment; (4) effectively only *Deutsche Physik* is denazified.

Naturally after the war the former followers of Lenard and Stark tried to defend themselves when attacked and to avoid part or all of the punishment entailed by denazification. The occupying powers for their part were most interested in the utilitarian value of German physics. The established physics community consciously created and consequently used *Deutsche Physik* as a scapegoat, for various reasons. Some scientists may have had suppressed feelings of guilt with regard to the colleagues they had not helped in 1933.[13] But by asserting that only *Deutsche Physik* had been politicized under National Socialism, the established physics community also killed several birds with one stone: (1) they appeared to be participating wholeheartedly in the denazification of their profession, to be putting their own house in order; (2) they avoided the denazification of the overwhelming majority of their colleagues; and (3) they tacitly asserted that their profession is inherently apolitical, thus a trustworthy servant worthy of generous support.

When the occupying powers called for denazification, it was in the context of denazification *and* demilitarization. However, similar to the case with denazification, it was not clear exactly what German demilitarization meant: was it the punishment of previous militarism, or the prevention of a revival of German militarism? John Gimbel's path-breaking work has demonstrated that the demilitarization of German science and engineering was fundamentally hypocritical,[14] indeed as demilitarization perhaps inherently is. German science — both eastern and western — was to be demilitarized, but individual German scientists and engineers were pressed or lured into the service of one of the victorious powers, in other words, pressed or lured into the service of someone else's militarism.

By the spring of 1945 Germany had lost a generation of physicists to the politicization of the universities and the carnage of the second world war. Physics also suffered during the destructive chaos at the end of the war and immediate postwar period. Those physicists who remained in Germany after the war were severely hindered in their research and harried by denazification.

However, many scientists avoided these restrictions by serving one of the four victorious powers.

"Specialist," the Soviet term for German scientists and engineers who traveled east after 1945 to work for the Soviet Union, aptly expresses how all four of the former allies viewed German scientists and engineers: as valuable and apolitical potential employees.[15] Andreas Heinemann-Grüder has argued persuasively that service for a victorious power — whether voluntary or not — retrospectively justified previous work for the Nazis and facilitated apologia.[16] Most important, German physics in both East and West were materially rebuilt by one of the two sides of the Cold War. By the fifties, German physics was a solid and well-integrated, if subordinate part of the international scientific community.

THREE ATOMIC BOMBS

A final comparison will hopefully shed light on the denazification and demilitarization of Germany physics: the American,[17] German,[18] and the Soviet atomic bombs.[19] There is a special German connection here: German émigrés helped build the American bomb, and German specialists helped build the Soviet weapon. With regard to the "hard" side of this comparison, scientists and engineers of comparable quality solved comparable problems, developed comparable technologies, and came to comparable results. However, one would expect that the motivations might be different for scientists working for a democratic United States, a fascist Germany, or a communist Soviet Union.

The German bomb project was driven, although certainly not as hard as its American or Soviet counterparts, by a desire to help Germany win the war, as well as personal and professional ambition. The American bomb was driven by a desire to help the United States win the war, as well as personal and professional ambition. The Soviet bomb was driven by a desire to help the Soviet Union win the Cold War, personal and professional ambition, and the unsubtle persuasion of the Soviet secret police, which was in charge of the project.

Thus the motivation of the researchers were not all that different, and hardly seem specific to their particular political or ideological system. All three groups of scientists were ambivalent about working on nuclear weapons, and this ambivalence helped to spawn respective, but different myths. In the United States, former members of the Manhattan Project, as well as some of their historians, place great emphasis on the Franck report, which opposed the use of the bomb in Japan, and the Oppenheimer Affair, when the scientist most responsible for the success of the American atomic bomb was unjustly stripped of his security clearance during the paranoia of McCarthyism, but do not like to be reminded of the fact that the bomb project which had begun because of the threat of Nazi nuclear weapons continued full blast after victory in Europe and that the first scruples were shown only after the successful test of a bomb in the New Mexican desert.

In the (former) Soviet Union, former members of the atomic bomb project place great emphasis on the patriotism which drove the project and the fact that the supposedly backward Soviet Union had achieved a nuclear weapon by 1949, but do not like to be reminded of the terrorizing presence of the Secret Police, Soviet scientific espionage, and the contribution of German specialists to their effort. In Germany, former members of the nuclear fission research project simultaneously asserted that their work had been of very high caliber and implied that moral scruples had been the main reason why nuclear weapons were never placed in Hitler's hands, but they do not like to be reminded of their often close collaboration with National Socialism.[20]

These three examples reveal a pattern. If nuclear weapons had been created (and perhaps used) and like in the United States and the Soviet Union one could plausibly argue that the cause had been just, then this just cause became the reason why atomic bombs were built. If like in Germany one could not plausibly argue that the cause had been just, but nuclear weapons fortunately had not been manufactured, then one asserted or implied that they had refrained from creating nuclear weapons in order to resist this unjust cause.

CONCLUSION

The Third Reich and postwar Allied occupation had a profound effect on German scientists' and engineers' self-image and postwar myths as well as on their science and technology. German physicists lived through nazification and denazification, militarization and demilitarization, by adapting to and working within the political and ideological system in which they found themselves. The one thing that did not change was their willingness to serve. The postwar victorious powers as well as the two new German states were in complete agreement with their scientists on one point: if physics was useful, and what is more useful than powerful new weaponry, then physicists would be used. Physicists were seen first and foremost as tools, and tools do not need to be nazified or militarized, denazified or demilitarized.

NOTES

1 Department of History, Union College, Schenectady, New York, 12308–2365 USA; a preliminary version of this article has appeared in Walter Kertz (ed.), *Hochschule und Nationalsozialismus. Projektberichte zur Geschichte der Carolo-Wilhelmina* (Braunschweig. TU Braunschweig, 1994), pp. 79–89.

2 However, see Herbert Mehrtens, "Kollaborationsverhältnisse: Natur- und Technikwissenschaften im NS-Staat und ihre Historie," in Christoph Meinel

and Peter Voswinkel (eds.), *Medizin, Naturwissenschaft, Technik und Nationalsozialismus — Kontinuitäten und Diskontinuitäten* (Stuttgart, GNT Verlag, 1994), pp. 13–32; as well as the articles in Monika Renneberg and Mark Walker (eds.), *Science, Technology and National Socialism* (Cambridge, Cambridge University Press, 1993) and some of the contributions to Josiane Olff-Nathan (ed.), *La Science sous le Troisième Reich*, (Paris, Le Seuil, 1993).

3 "*Deutsche Physik*" is literally translated as "German Physics" and often as "Aryan Physics"; for *Deutsche Physik* see Steffen Richter, "Die Kämpfe innerhalb der Physik in Deutschland nach dem 1. Weltkrieg," *Sudhoffs Archiv* (1973) *57*: 195–207; Alan Beyerchen, *Scientists under Hitler: Politics and the Physics Community in the Third Reich* (New Haven, Yale University Press, 1977); Steffen Richter, "Physiker im Dritten Reich," *Jahrbuch der TH Darmstadt* (1978/1979), 103–13; Andreas Kleinert, "Von der Science allemande zur Deutschen Physik: Nationalismus und moderne Naturwissenschaft in Frankreich und Deutschland zwischen 1914 und 1940," *Francia* (1978) *6*: 509–25; Andreas Kleinert, "Nationalistische und antisemitische Ressentiments von Wissenschaftlern gegen Einstein," in H. Nelkowski *et al.* (eds.), *Einstein Symposium Berlin* (Berlin, Springer, 1979), pp. 501– 516; Steffen Richter, "Die 'Deutsche Physik,'" in Herbert Mehrtens and Steffen Richter (eds.), *Naturwissenschaft, Technik und Ideologie: Beiträge zur Wissenschaftsgeschichte des Dritten Reiches* (Frankfurt a.M., Suhrkamp, 1980), pp. 116–41; Andreas Kleinert, "Das Spruchkammerverfahren gegen Johannes Stark," *Sudhoffs Archiv* (1983) *67*: 13–24; Andreas Kleinert, "La correspondance entre Philipp Lenard et Johannes Stark," in Olff-Nathan, pp. 149–66; for physics during the Third Reich also see Beyerchen; Monika Renneberg, "Zur Mathematisch-Naturwissenschaftlichen Fakultät der Hamburger Universität im 'Dritten Reich,'" in Eckart Krause, Ludwig Huber, and Holger Fischer (eds.), *Hochschulalltag im "Dritten Reich," Die Hamburger Universität 1933–1945* (Berlin, Reimer, 1991), pp. 1051–1071; Renneberg, "Die Physik und die physikalischen Institute an der Hamburger Universität im 'Dritten Reich,'" in Krause, *et al.*, pp. 1097–1118; David Cassidy, *Uncertainty. The Life and Science of Werner Heisenberg* (New York, Freeman, 1991); Michael Eckert, *Die Atomphysiker* (Braunschweig, Vieweg, 1993); M. Norton Wise, "Pascual Jordan: Quantum Mechanics, Psychology, National Socialism," in Renneberg and Walker, pp. 224–54, 391–396; Klaus Hentschel and Monika Renneberg, "'Ausschaltung' oder 'Verteidigung' der allgemeinen Relativitätstheorie — Interpretationen einer Kosmologen-Karriere im Nationalsozialismus," in Meinel und Voswinkel, pp. 201–207; Mark Walker, *Nazi Science: Myth, Truth, and the German Atomic Bomb* (New York, Plenum Press, 1995).

4 See for example Klaus Hentschel, *Interpretationen und Fehlinterpretationen der speziellen und allgemeinen Relativitätstheorie durch Zeitgenossen Albert Einsteins* (Basel, Birkhäuser, 1990).

5 Monika Renneberg und Mark Walker, "Scientists, Engineers, and National Socialism," in Renneberg und Walker, pp. 1–29, 339–46.

6 For Stark during the Third Reich, see Walker, *Nazi Science*, chapters 2 and 3.

7 See the articles in Renneberg and Walker.

8 Cassidy, 346–414.

9 See Clemens Vollnhals, *Entnazifizierung. Politische Säuberung und Rehabilitierung in den vier Besatzungszonen 1945–1949* (Munich, DTV Verlag, 1991).

10 Lutz Niethammer, *Die Mitläuferfabrik. Die Entnazifizierung am Beispiel Bayerns* (Berlin, 1982 2).

11 Vollnhals, p. 57, "Ohne die Zustimmung und Mitarbeit der Deutschen selbst liess sich eine politische Säuberung wohl administrative von oben verordnen, nicht aber effektiv und dauerhaft wirksam durchführen".

12 See Giles' contribution to this volume.

13 Thanks for Prof. Albrecht, Technical University of Braunschweig, for this observation.

14 John Gimbel, *Science, Technology, and Reparations. Exploitation and Plunder in Postwar Germany* (Stanford, 1990).

15 Ulrich Albrecht, Andreas Heinemann-Grüder and Arend Wellmann, *Die Spezialisten. Deutsche Naturwissenschaftler und Techniker in der Sowjetunion nach 1945* (Berlin, 1991)

16 Andreas Heinemann-Grüder, "*Keinerlei Untergang*: German Armaments Engineers during the Second World War and in the Service of the Victorious Powers," in Renneberg und Walker, pp. 30–50, 346–52.

17 Richard Rhodes, *The Making of the Atomic Bomb* (New York, 1986); Lillian Hoddeson, Paul W. Henriksen, Roger A. Meade, and Catherine Westfall, *Critical Assembly: A Technical History of Los Alamos during the Oppenheimer Years, 1943–1945* (Cambridge, Cambridge University Press, 1993).

18 Mark Walker, *German National Socialism and the Quest for Nuclear Power, 1939–1949* (Cambridge, Cambridge University Press, 1989); Walker, *Nazi Science*.

19 Andreas Heinemann-Grüder, *Die Sowjetische Atombombe* (Münster, Verlag Westfälisches Dampfboot, 1992); David Holloway, *Stalin and the Bomb: The Soviet Union and Atomic Energy, 1939–1956* (New Haven, Yale University Press, 1994).

20 Thus I reject attempts by Thomas Powers and others to revive Robert Jungk's 1955 thesis of a conspiracy among German scientists to deny nuclear weapons to Hitler; see Thomas Powers, *Heisenberg's War* (New York, Knopf, 1993); the recently published "Farm Hall Transcripts" provide additional evidence which refutes this claim of a conspiracy, see *Operation Epsilon: The Farm Hall Transcripts* (Bristol, Institute of Physics Publishing, 1993); Robert Jungk has recently distanced himself from his original thesis; see his memoirs, *Trotzdem. Mein Leben für die Zukunft* (Munich, Hanser, 1993), pp. 298–300.

CHAPTER 5

Denazifying Scientists — and Science

Mitchell G. Ash

In recent years a transformation has occurred in our historical understanding of science and technology under Nazism, with important implications for the postwar period. Evidence is mounting that so-called "normal" science and technology not only survived, but prospered under Nazism, particularly in fields perceived to be closest to its core projects — world conquest and the creation of a new, dominant German race.[1] In these projects internationally accepted science and technology lived side by side with pseudoscience sponsored by the SS and other organizations. Both could be supported by the Nazi state, because its polycratic structure offered scientists and engineers multiple patrons to whom they appealed.[2] In mobilizing modern science and technology for its purposes, the Nazi state was not entirely parasitic on innovations from the past, as Hans Mommsen claims that it was in other fields.[3] The fact that the Allies showed such interest in some of the results is evidence for the claim that in disciplines like aeronautics and rocketry, for example, innovations took place with state support during the Nazi period that cannot easily be dismissed as pseudoscience.

In light of this revised appraisal of science and technology under Nazism, the denazification question becomes more complex — and intellectually interesting — than before. The assumption on which the exploitation and transfer of science and technology by the Allies rested was a view that I would like to call "technocratic innocence" — the belief that science and technology are essentially apolitical, value-neutral instruments or tools, easily transferrable cultural capital. This assumption is questionable, because it ignores the fact that scientists actively attached themselves rhetorically or more concretely to core projects of Nazism, thus undermining the claim that science is *per se* value-neutral by renegotiating the relationships between science, technology and politics. Moreover, this view overlooks the fact that science and technology are not closed and static, but open and dynamic systems, which means that both scientific "products" and the use value of their producers constantly change. Thus, for the postwar period, the question becomes not which Allied or postwar German regime gathered in the largest amounts of available human and scientific resources, but rather, the ways in which both the availability of scientists and the relationships of science, technology and politics changed in different political and social circumstances.

The criteria employed by policy makers and the strategies followed by scientists during denazification did not differ as completely on the two sides of the Iron Curtain as is often assumed. Both Communist and non-Communist policy makers faced tensions between broadly conceived political aims and the need for specialists to support economic reconstruction. As a result, significant numbers of scientists who had served the Nazi state survived denazification, negotiated new careers, reconstructed their research programmes and realigned those programmes directly or indirectly with political or technocratic projects, even in the G.D.R. Ironically, precisely their successes in constructing continuity with the past often blocked innovation after 1945.

DENAZIFICATION POLICIES AND THE AVAILABILITY OF SCIENTISTS

Although examples from the universities are often cited in general accounts of denazification, full-scale comparative analysis of denazification policy in higher education and the scientific sphere has only begun.[4] General conclusions will probably be limited for some time by the difficulty of accessing key records, particularly those of denazification proceedings in the Western zones as well as CIA and State Department files and the records of the Soviet military government in Moscow.[5] Nonetheless, two general points can and should be made at the outset.

1. With the important exception of Allied Control Authority Law 25 of 29 April 1946, which forbade both basic and applied research that could have military applications, science *per se* was not a central concern for occupation officers. Instead, policies enacted with larger political or economic ends in view affected scientific personnel alongside others. Denazification in higher education, for example, was driven less by the desire to exploit scientific and technical resources than by the perceived need to assure security and pave the way for "democratization" or "reeducation," however defined, by removing persons who had compromised themselves politically from teaching positions.

2. The Allies' exploitation and evacuation programs did not run parallel to all aspects of the denazification process, but preceded and undercut it at specific points. Though the timing differed in each zone, four phases can be distinguished. The first phase ran from occupation to the Potsdam conference (17 July–2 August 1945) and combined outright plunder of universities, Kaiser Wilhelm institutes and industrial laboratories with organized detention and evacuation of selected scientists and dismantling or transfer of equipment.[6] The

second stage, actually the initial phase of denazification proper, ran roughly from autumn 1945 to spring 1946. Dominant here in all four zones were complex pulling and tugging among Allied officers and German university, science and government administrators. As will be shown below, in these maneuverings university autonomy was as much at stake as the removal of Nazi scientists; summary dismissals by the Allies and more or less voluntary "self-cleansing" operations were both a part of the game.

During the third stage, beginning in the American zone with the "Law for the Liberation from National Socialism and Militarism" promulgated in March 1946 and some months later in the British zone, primary responsibility for denazification was gradually transferred to German administrators, with occupation officers taking supervisory roles (for differences in the French and Soviet zones, see below). The Allies' largest exploitation and evacuation operations (e.g., British T-force operations, FIAT, the massive Soviet evacuation of specialists in October, 1946 and, somewhat later, Project Paperclip) occurred primarily in this period. Allied evacuations of scientists thus undercut denazification precisely by circumventing the German officials now held reponsible for it.[7] It is quite possible that the resulting complications helped usher in stage four, the formal end of denazification in the Soviet zone with Order Nr. 201 of August 20, 1947, soon followed by comparable measures in the Western zones. This signaled the gradual return of many so-called "nominal" NSDAP members or "collaborators" (*Mitläufer*) to academic and scientific life.

The following brief comparative account — limited for practical reasons primarily to universities and technical academies — brings out the similarities and differences in policies and their impacts in the four occupation zones.

The Western zones

It is generally agreed that the Americans were more firm and schematic and less pragmatic in their approach to denazification than the French and British. According to an intelligence report of 1947, a major purpose of American policy in higher education was to give Germans "practice" in democracy by letting them have as much of a role as possible in university reform, while university officers and their superiors set broad guidelines and retained ultimate control over denazification of personnel.[8] In practice, denazification was often treated as a security issue and handled by Special Branch or Counter-Intelligence Corps (CIC) officers, whose harsh views frequently conflicted with those of university officers.[9]

Recent local studies and data compiled at the time show that pulling and tugging between German public and university officials and American university officers during the initial stage of denazification resulted in the dismissal of hundreds of academics from the universities in Frankfurt, Heidelberg, Marburg, Munich, and Erlangen between 1945 and 1947, with most dismissals coming relatively early in the occupation.[10] James Tent estimates that in fall 1946 and early 1947 "approximately fifteen per cent" of these already-pruned faculties were removed in a second purge in response to press reports in America that denazification was not rigorous enough.[11] The problem, in the end, was that many of these people refused to stay fired. Either they moved to universities in the other Western zones or took advantage of the shift of the primary burden to German officials, which obscured the distinction between political purge and rehabilitation, and continued arguing their cases until they ultimately returned to academic life in the 1950s.[12]

In the British zone, a policy that was intended to be in deliberate contrast to the rigorism of the Americans produced a similar result. Former science officer Colonel Bertie Blount recalls that "witch-hunting" was not British policy. Rather than requesting or reacting to advice from different individuals on a case-by-case basis, as the Americans did, the British military government in Minden created scientific advisory groups that provided extensive advice to occupation officials.[13] In higher education, denazification was seen primarily as a security issue and assigned to the Public Safety Special Branch. British education officers later recalled that they were glad that the job belonged to someone else.[14] But the priority that the British assigned to economic reconstruction and the corresponding need for qualified experts made a rigorous purge impracticable. The relocation of many of the best German scientists to the British zone had much to do with this general outlook.[15]

Actual administration of what policy there was in the British zone was inconsistent and inefficient. One result was that Fritz Lenz, former head of the Department of Eugenics of the Kaiser-Wilhelm-Institute for Anthropology in Berlin, who had pointed to affinities between the views of the race hygiene movement and those expressed in *Mein Kampf* as early as 1931, became a professor in Göttingen in 1946 with no apparent difficulty. A recent study of five institutions in Lower Saxony shows that by July, 1947, 31 per cent of all professors, Dozenten and assistants (208 of 676) who had been on staff in 1945 had been dismissed; 25 per cent of these (51) appealed successfully, leaving a total of 23 per cent (157) dismissals. No particular favoritism was accorded to science and technology; the Technical Academy in Braunschweig had the highest dismissal rate (49 per cent), while its counterpart in Hannover had the lowest (15 per cent). However, there was no net loss in the end, because 240 new appointments more than made up for the dismissals. After 1947, many who had been dismissed eventually returned. Here, as in the American zone, professors persistently and skillfully pursued their reinstatement, and the majority of their colleagues proved willing to take them back.[16]

Denazification policy in the French zone was widely regarded to be far more lenient than that of the Americans, or even the British; but the results appear to have been much the same. After an initial period of dismantling, French science policy emphasized attempts to attract German scientists, including former Nazis, integrate them into French or German universities, and rebuild their institutes with a combined aim of cooperation and control. A declaration of April 1946 explicitly stated that denazification should not be reduced to a political purge.[17] Thus, after the removal and transport to Paris of the high-tension apparatus from the Kaiser Wilhelm Institute for Chemistry in May 1945, the institute resumed work without restriction under the "protection" of the "Second d'Information Scientifique."[18] In higher education, teachers were either permanently dismissed, retained or provisionally dismissed and then readmitted after examination of their files. As of April 1, 1946, 101 university teachers (26.3 per cent) had been removed and 283 (73.7 per cent) retained or readmitted — a figure comparable with those just cited for the British zone.[19] In addition, the French tried to change the political mix by founding new universities in Mainz and Trier and involving themselves actively in their staffing, for example by importing French faculty and assistants and recruiting other faculty from outside academic circles. The French officer responsible for this sector declared in early 1947 that 25 per cent of the successful applicants for positions in Mainz had been "nominal" NSDAP members.[20]

The Soviet zone

In the Soviet zone, the general tension between an ideological-political goal and its pragmatic, often inconsistent, execution was comparable with that in the Western zones.[21] Here the actual aim — never openly stated — was not "reeducation" for liberal democracy, but eventual restructuring of scientific and higher educational institutions under Communist (later Socialist Unity) Party tutelage. Once the Soviet military government (SMAD) decided to open the universities, however, battles immediately began between university faculty officials and the Science and Higher Education Department of the Deutsche Verwaltung for Volksbildung (DVV), established by the Soviets in August, 1945. As in the Western zones, university autonomy was as much at stake as the removal of Nazi scientists. In Berlin, the university and other scientific institutions in the Soviet sector were placed directly under DVV jurisdiction in September 1945.[22] Elsewhere, the struggle for control was complicated by tensions between the DVV and the newly-created *Land* and provincial administrations, which led to considerable variation in the effectiveness of denazification (see below). In addition, there were abrupt course changes by the SMAD itself, which sometimes accused the DVV of being too strict and sometimes of being too lenient, and always demanded detailed responses with data in two weeks or less.[23]

The lines of conflict and confusion emerged quickly. At the Mining Academy in Freiberg, for example, a provisional academic senate quickly dismissed the most active Nazi faculty, and in November 1945, Saxony's Department of Education ordered the firing of all former Party members. Both times local Soviet occupation officials intervened, the second time going to the Supreme Command of the SMAD, to assure that all scientists would have the chance to remain, if not to teach, then at least to work on Soviet research projects.[24] In an October, 1945 meeting with DVV President Paul Wandel, Leipzig University rector Bernhard Schweitzer, a theologian, accused Saxony's Education

Department of dictatorially presenting a list of university teachers to be fired at the last minute, and claimed that the local Soviet military commandant had said, "please do not purge too much, or you will have no professors left." Wandel then agreed that former NSDAP members might be retained who were irreplacible in teaching (especially in medicine), or to prevent elimination of a subject from the curriculum. Thus, even at this early stage pragmatic criteria began to play a role in the discussion.[25]

However, such moves to resolve the tension between ideology and pragmatism did not go far enough. The request by Schweitzer's successor, philosopher Hans Georg Gadamer, to close the Leipzig medical faculty in July 1947, may have played a role in the SMAD's move to integrate "nominal" Nazi party members in all walks of life in its Order Number 201 of August 1947.[26] Despite repeated calls by the Socialist Unity Party (SED) to break the power of Germany's traditional elites, the SMAD was aware that scientific expertise, once lost, would be difficult to replace, and the SED itself needed to retain "bourgeois" experts until a "new intelligentsia" could take their places. Thus, in striking contrast to the schematic dismissals of the vast majority of schoolteachers in the Soviet zone, the DVV insisted on a case by case assessment of professors, including their writings as well as their functions in Nazi organizations.[27]

For the first two years of Soviet occupation, however, there was no uniform procedure for carrying out such assessments. The DVV personnel department assembled files from the Reich Education Ministry and the National Socialist Professors' League for more than 3,000 German university teachers.[28] But it did not create a separate section for universities and science until 1947; a nearly complete card file for all university teachers in the Soviet zone was not available until 1948.[29] Systematic political assessments thus become routine only after the end of denazification as such was on the horizon. By that time — and still more so after the founding of the G.D.R. in 1949 — the aim was clearly the controlled integration rather than the elimination of former Nazis.

In the initial wave of dismissals (fall 1945 — spring 1946), 427 faculty in Berlin (78 per cent), 138 (73 per cent) in Greifswald, 53 (67 per cent) in Dresden, 165 (67 per cent) in Leipzig, 67 (55 per cent) in Rostock, and 98 (46 per cent) in Jena lost their posts.[30] These dismissal rates were far higher than in any of the Western zones, despite the wide range of variation within the Soviet zone. But even in this part of Germany a pattern of creeping reinstatement set in almost immediately. In 1948 a total of 747 professors and Dozenten were working in universities and technical academies in the Soviet zone, a drop of 58.9 per cent compared with 1819 on the same territory in 1938; of these 90 (12 per cent) were former NSDAP members, 24 appointed in 1946, 39 in 1947, and 27 in 1948.[31] By 1954 there were 868 professors and Dozenten in the G.D.R., and the percentage of former NSDAP members among professors alone (28.4) was roughly equal to that of SED members (28.7). The levels were higher still in the natural sciences (31.2 per cent), technical fields (41.9 per cent) and medicine (45.9 per cent). Ralph Jessen understates the case when he writes that "a brown past was certainly not an insupportable career barrier in the higher education institutions of the G.D.R. in the 1950s."[32]

Summary. The enormously high turnover of academic and scientific staff in this period — 90 per cent of all professors and Dozenten listed at the University of Berlin alone between 1944 and 1948, for example — cannot be attributed solely to dismissals under denazification and Allied evacuations.[33] The number of scientists arrested by the NKVD and either deported to the Soviet Union or interned in camps in the Soviet zone is not yet known. More biographical information will be needed in order to measure all the factors involved directly.

Thus, only a very general statement about the relative short-run impact of denazification on the availability of researchers can be made at this stage. A 1950 accounting by the journal *Christ und Welt* states that 4,298 academics were dismissed after 1945. Of the 3,479 of these who were presumed to be in Germany in May, 1950, 1,301 (37.4 per cent) had been rehabilitated.[34] Unfortunately, this source — like the other studies cited above — gives no breakdown by discipline, and does not count personnel of Kaiser Wilhelm Institutes or Wehrmacht and industrial laboratories without academic appointments. Nonetheless, the total of 2,178 unreinstated academics in all fields in 1950 is far higher than the number of specialists brought to the United States under Operation Paperclip (a total of 555 by 1952), and slightly less than the 2,370 German specialists brought to the Soviet Union.[35]

The number of reinstatements climbed rapidly after the Federal Republic's Basic Law restored civil servants to their former status in 1949. Despite official mythologies to the contrary, significant numbers of former NSDAP members were also tolerated in the G.D.R. Who the former Nazis were and why they were or were not tolerated in the different zones at specific times is suggested by the following case studies.

DENAZIFYING SCIENCE: CASE STUDIES FROM GENETICS AND PSYCHOLOGY

At work in all of these cases was a complex interplay of individual self-interest, networking among colleagues, and changing political circumstances. The most important common thread was the instrumentalization of basic research — its association, not with large pieces of apparatus, but with technological or technocratic projects. The following West and East German cases from genetics and psychology show how such linkages made it possible for scientists who had collaborated directly or implicitly with Nazi policies before 1945 to advance their careers afterward, whether or not they had belonged to the NSDAP.

Genetics 1: Otmar von Verschuer

The case of human geneticist and eugenicist Otmar von Verschuer exemplifies the relative severity and legalism of denazification policy in the American zone in the early stages of the occupation. At the same time, it shows how dependent the Americans were on the willingness of colleagues to come forward rather than protect their own. This they did here for a mix of political and professional reasons.

Verschuer was one of the most prominent younger members of the eugenics movement in Germany. A medical scientist who had pioneered the use of blood group diagnosis and statistical methods in twin research, he became head of the Department of Human Heredity at the Kaiser Wilhelm Institute for Anthropology, Human Heredity and Eugenics when it was founded by Eugen Fischer in 1927. He joined the NSDAP in 1940 and actively supported eugenical measures, for example by serving on "hereditary health courts" created to decide compulsory sterilizations. Called to Frankfurt to head a newly founded Institute for Biology and Race Hygiene in 1935, he returned to Berlin as Fischer's successor in 1943. In 1943 and 1944 he was nominal director of a "research project" carried out by his former assistant Joseph Mengele while the latter was at Auschwitz, and Mengele sent blood and organ samples to Berlin for analysis. The circumstances of that collaboration, which did more than anything else to blacken the name of eugenics, came to light during Verschuer's denazification proceedings.

In spring 1945, while Verschuer was negotiating with Frankfurt officials about reappointment to his former professorship there, counter-pressure came from the American military government in Berlin, with help from a Communist. Robert Havemann, a physical chemist and resistance member who had been named "President" of the Kaiser Wilhelm Society (KWG) by the Berlin Magistrat, wrote to American Education and Religious Affairs (E and RA) department officer Samuel Shulits on 18 February 1946 calling for the return of the Institute's property of Dahlem, informing him of Verschuer's link to Mengele and placing him "among the prominent and most active representatives of fascism among scientists ..." The affair became public when badly garbled newspaper reports of interviews with Havemann appeared in both the American and Soviet sectors of Berlin in May.[36]

Because Verschuer had evacuated most of his institute's inventory to a Hessian village where his family had an estate, sociologist Edward Hartshorne, E and RA officer in Hesse, investigated in June 1946. In July he reported that, "On the face of it Verschuer appears to be unusable under present directives." Verschuer visited Hartshorne in Frankfurt 8 July, and the Institute's inventory was moved to Frankfurt two weeks later. But on 24 July two American intelligence officers took his papers and put him under house arrest until further notice. Hartshorne's widow later reported that her husband wanted the Institute brought to Frankfurt, but never intended to allow Verschuer to continue as director.[37]

Verschuer nonetheless pressed ahead with denazification proceedings, partly to clear his name and partly because he thought that his appointment depended on it. In a questionaire dated 23 April 1946 he had admitted membership in the NSDAP, the NS Physicians' League and the NS Welfare Organisation, but described himself as "a cooperator or innocent" (*Mitläufer oder entlastet*) in part because of his parallel affiliation with Pastor Martin Niemöller's Confessing

Church in Berlin-Dahlem. A city official wrote to the Hessian Ministry in charge on 2 November 1946 requesting rapid action because of the impending appointment. One week later Verschuer was classified as a *Mitläufer* and fined 600 Marks.[38]

Contrary to widespread accounts, however, the story did not end there; here, at least, the *Mitläuferfabrik* did not function smoothly. Havemann filed a formal protest against this decision on 23 January 1947, and Hesse's Minister of Culture ordered Verschuer's removal on 28 February.[39] An American Special Branch liason wrote to the public prosecutor on 14 May that he would ask the denazification authority to vacate its decision and order a new trial. But the proceedings against Verschuer were not reopened until 8 March 1949, by which time denazification in the American zone had long since ended. Thus, American action considerably delayed, but could not prevent his eventual reinstatement, albeit in Münster, a university in the more lenient British zone.[40]

Verschuer was named full professor in Münster in 1951, and soon became the leading scientist in West Germany in the field of human genetics. Just at the time when molecular genetics had its breakthrough with the work of Watson and Crick, and the so-called "modern synthesis" of population genetics and evolutionary theory had become the predominant paradigm in non-molecular biology, Verschuer and his coworkers continued using the research techniques that had been standard in eugenical human genetics before 1933 — statistical assessment of geneological survey data and twin studies. In the 1950s, he and his students added a longitudinal dimension by following up on twins he had studied earlier in Tübingen, Frankfurt and Berlin.[41] An American reviewer considered this work important due to the scarcity of longitudinal twin studies, but noted that the work was methodologically out of step — "no objective psychological tests were employed, and only the simplest kinds of statistics were used for interpreting the data."[42]

A dimension of change came into the picture when Verschuer reconstructed at least one part of older eugenical research practice, albeit without state-mandated compulsory sterilization. In response to world-wide concern about the effects of radiation on mutations, he established extensive heredity registers in Westphalia quite similar to those he had set up for eugenical purposes in Frankfurt during the 1930s, with generous support from the city of Münster and the government of North-Rhine-Westphalia, thus re-legitimating earlier methods in adaptation to a current trend. The date turned out to be worthless for studying the effects of radiation on mutations. Modern molecular and population genetics only took hold in West Germany after Verschuer's ungentle removal from the leadership of the field on his retirement in the early 1960s.[43]

Genetics 2: Hans Stubbe and the Institute for Domesticated Plant Research

The case of plant geneticist Hans Stubbe's Institute for Domesticated Plant Research indicates circumstances in which authorities in the Soviet zone and the G.D.R. not only tolerated, but even provided extraordinary support for former Nazi party members. From the late 1920s Stubbe had worked at the KWI for Breeding Research, headed by leading plant geneticist and eugenicist Erwin Baur. Working, as did Baur, with the snapdragon (*Antirrhinum*), Stubbe studied induction of mutations by chemicals and radiation as well as effects of other factors on mutation rates, all work on the forefront of modern genetics. He had belonged to the SPD before 1933, and never joined the NSDAP. Fired in a dispute with the acting director of the institute in 1936, and unable to get an academic job because he was deemed politically "unreliable," Stubbe went to the KWI for Biology, headed by Fritz von Wettstein, later portrayed by many as a "spiritual oasis." He founded the Institute for Domesticated Plant Research (*Kulturpflanzenforschung*) in Austria with Wettstein's support in 1943. His radiation genetics studies received more funding than the work of any other junior scientist in the KWG from 1934 to 1945.[44]

After the war, Stubbe was able to move without difficulty to the Soviet zone and reestablish his institute in the village of Gattersleben, near Halle. He was appointed professor at the university in Halle, and later also became head of an academy attached to the Ministry of Agriculture in East Berlin. His political credentials were evidently sound enough for him to be appointed in 1945 to a committee in charge of land reform, and the regional Soviet commandant also entrusted him with the direction of three local seed companies.[45] The fascination of radiation biotechnology and its promise for agriculture were more than enough to earn Stubbe and his colleague Hans Mothes the Nationalpreis of the G.D.R. in 1949, the first year of its existence. A 1955 report shows steady growth from 10 scientists and 66 support staff in 1948, when it was attached to the Academy of Sciences, to 34 scientists and 255 staff in 1952.[46]

This meteoric rise was threatened by the Lysenko controversy, at the height of which Stubbe received an offer from the Max Planck Society (MPG). As a result he was able to get his complaints about budget reductions and the exclusion of his genetics department from the first five-year-plan heard by Walter Ulbricht himself in February 1953. He openly reported to Academy of Sciences administrator and SED official Joseph Naas in June 1952 "that ideological questions of genetics have no influence on the acceptance or rejection of the offer from the MPG," since three years of experiments to test Lysenko's claims had all had negative outcomes.[47]

Stubbe later came to be regarded as something of a hero for his resistence to Lysenko, and in his case even the Central Committee of the SED had to alter its attitude. In a 1953 report, officials there hoped to make him state his views publically in order to provoke a controversy that could be used for propaganda purposes. A later report on the political and social status of the Institute's personnel, possibly assembled to keep such data in readiness for an eventual purge, stated that none of the four leading scientists belonged to any party in 1954, but three — i.e., all but Stubbe himself — were former NSDAP members; of the 26 scientific associates, 6 were former Nazis. Of the 539 staff for whom party affiliation was documented, a total of 44 were former Nazis.[48] In 1956, however, after the fall of both Stalin and Lysenko, the Central Committee was forced to acknowledge that "the national and international significance and the scientific achievement" of the Institute had not been "correctly evaluated" earlier. Recommended now were improvements in supplies to the town of Gattersleben, a clinic, apartments for institute personnel, more equipment, more rapid import of Western literature, and even a hunting rifle for one of the department heads to battle varmints on the experimental fields.[49]

A 1957 report of the G.D.R. Academy of Sciences proudly proclaimed that Stubbe's institute "has influenced the economic and cultural development of the G.D.R. in many ways," and promised that mutation research in the genetics department "will have a positive impact on our agriculture in the near future by making available valuable new varieties (especially of summer and winter wheat, soybeans and tomatoes)."[50] It would appear that, in this case, continuity with technocratic paradigms and research styles from the past did not mean falling out of step with international trends, as it did in the case of Verschuer.

Psychology 1: Oswald Kroh

The case of Oswald Kroh indicates quite clearly that the tension between ambitious political goals and realities on the ground was as much a part of denazification in the Soviet zone as it was in the Western zones. An educationalist and developmental psychologist, Kroh became one of Germany's leading psychologists under Nazism by effectively combining ideology and technocracy. A German nationalist before 1933, he joined the NSDAP in that year. By 1935 he could proudly state that he was on the editorial board of *Die Rasse* and "Liason for the German Society for Psychology to the Delegate of the Führer for the Entire Intellectual and World-View Education of the NSDAP."[51] The ideological vehicle for all this was his support for the "folkish education" (*völkische Erziehung*) advocated by Ernst Krieck.[52] He moved from Tübingen to a more prestigious chair in Munich in 1938, with the warm recommendation of the local head of the Nazi Professors' League, and ultimately became professor of philosophy and head of the Berlin psychological institute, the most prestigious in Germany, in 1942.

Since 1940 Kroh was also chair of the German Society for Psychology, and in that capacity he actively promoted the professionalisation of the discipline on the basis of its applications in personnel selection, especially for the Wehrmacht. Although military psychologists had no use for the vague categories of "race psychology" or "folkish education," Kroh saw no conflict between his commitment to the discipline's professional advancement and his version of Nazi ideology.[53]

Kroh was dismissed from his Berlin position on 12 December 1945. While he struggled to rehabilitate himself, he was kept going for two years with low-paid research stipends — one of two dozen former Nazis supported by this means at the University of Berlin. In his rehabilitation efforts, he mobilized support from a wide range of colleagues, including known antifascists such as Heinrich Düker, respected non-Party colleagues such as Ernst Kretschmer, and

J.E. Heyde, a folkish educationalist like himself who had in the meantime been appointed to a professorship in Rostock.[54] In pedagogical congresses called in the Soviet zone, he was well received — to the consternation of observers from the DVV.[55] As a result, Robert Heise, dean of newly established Pedagogical Faculty at the Humboldt University, proposed him for a professorship of psychology.

This led to a split among ideologists and pragmatists within the DVV along lines similar to corresponding discussions in the Western zones. Juvenile delinquency was a major social problem in Berlin, and Heise characterized Kroh as a much-needed expert in youth affairs. DVV Vice-President Marquardt noted that Kroh was being consulted by the DVV itself in the development of assessment questionaires for school children, and pointed to the precedent of Heyde's appointment in Rostock. In contrast, Willi Lehmann, head of the DVV personnel department, called Kroh "a gross questionaire fraud (*Fragebogenfälscher*) and long-time loyal servant of Nazism," and concluded that a university appointment for him was "out of the question."[56]

In June 1948 DVV President Paul Wandel pressed for a decision, to be based not only on Kroh politics, but also on "what sort of psychology instruction" should be conducted in the Pedagogical Faculty. After considering other information, including documentation of Kroh's participation in a "Germanization" program (*Eindeutschungsprogramm*) near Posen in 1942, DVV Vice President for Science and Higher Education Robert Rompe concluded, apparently reluctantly, that his appointment "would not be understood" by teachers.[57]

By this time Kroh had already applied for the directorship of the Psychological Institute at the newly-founded Free University in Berlin's American sector. A denazification proceeding and a public campaign against him delayed the appointment until 1950, but then he held the position until his death in 1956. There he continued to propagate the holistic conceptions of personality and intuitionistic methods of personnel diagnostics that had predominated in German psychology during the 1930s. Not until the early 1960s did the American-style statistical approaches to personality assessment advocated by younger psychologists take precedence.[58] Thus, in this discipline, as in genetics, a shift in scientific orientation that might have happened sooner in West Germany was actually delayed in part by the continued employment of former Nazis.

Psychology 2: Kurt Gottschaldt

A strong contrast to Kroh's odyssey is that of Kurt Gottschaldt.[59] From 1935 to 1945, Gottschaldt headed the Department of Hereditary Psychology at the Kaiser Wilhelm Institute for Anthropology; he was appointed associate professor at the University of Berlin in 1938. His major achievement in the Nazi period was the creation of so-called "twin camps" for long-term observation of twins in the summers of 1937 and 1938, with funding from the NSV Gau Berlin and the Interior Ministry. The central issue of his research was the heritability of behavior and personality characteristics, assessed with a combination of phenomenological observation, social-psychological experiments and statistical comparisons among twin pairs. Gottschaldt never joined the NSDAP and voiced open commitments neither to Nazism's aims nor even to eugenical practices. It was institute director Eugen Fischer who emphasized the importance of Gottschaldt's work for "positive racial hygiene." His research was also supported by Fischer's successor, Otmar von Verschuer.[60]

Gottschaldt joined Verschuer when the Institute was transferred to Hesse in February, 1945, but returned to Berlin with his superior's permission in August. In May 1946 he enlisted both American E and RA officer Samuel Shulits and Robert Havemann in an effort to recover his twin research data from Verschuer (see above).[61] Both KWI for Biophysics head Boris Rajewsky and Verschuer himself clearly believed that it was Gottschaldt who began the investigation that led to the discovery of Verschuer's connection with Mengele and Auschwitz.[62] There is no doubt that he was the chief informant of both Robert Havemann and of American war crimes investigators in this matter.[63] Thus, the denunciation of Verschuer and the exposure of the Mengele connection was not only an act of political courage. Since he was not a geneticist, Gottschaldt did not belong to the inner circle of colleagues in that field who protected one another. Lacking allies in his own field of psychology, he was dependent on his wits and on alternative alliances, in this case with Robert Havemann and the Americans — a combination that was possible only in 1946, while the future of Berlin was still unclear.

In January 1946, Gottschaldt had assumed the equivalent of his former academic position by becoming associate professor (*Professor mit Lehrauftrag*) in the Mathematics and Natural Sciences Faculty of the University of Berlin.[64] Soon he was proposed for a full professorship. Opinions solicited by the faculty on the issue were mixed, but none opposed him on the basis of his scientific support for positive eugenics under Nazism. Testimonials came primarily from colleagues in the Western zones, including his pre-1933 superior in Bonn, Erich Rothacker. Apparently no one noticed or cared about Rothacker's earlier active membership in the NSDAP. After some delay, Gottschaldt was named by the DVV itself without a faculty vote on 1 December 1947. The move was based both on practical considerations — the need to establish instruction in psychology in the Mathematics and Natural Sciences Faculty — and on explicit support for his natural scientific approach to psychology.[65] This case and that of Oswald Kroh, just discussed, are concrete indications of the wide range of behavior that was covered by the DVV's policy of individual assessment of prospective university teachers.

In the G.D.R., Gottschaldt continued the twin research that he had begun under Nazism with only slight alterations. Indeed, in the early 1950s he applied for and received funding from both German governments for a follow-up study of the twins he had investigated in the 1930s.[66] Neither of his funding applications contained obvious ideological appeals, but this was not a sign of value-neutrality. Both regimes had obvious interest in the issue of heredity and its limits; the question's relevance both for educational policy and for personnel diagnostics seemed obvious at the time.

An indication of change in Gottschaldt's thinking while he worked in the G.D.R. was a 1954 reference to the importance of socioeconomic determinants of personality as well as hereditary "constitution."[67] Also new was the concept of social "superstructure" (*sozialer Überbau*) of the person, which in his view begins to appear in puberty. Though such ideas were potentially relevant to socialist pedagogy, that relevance was not explicitly stated. It was not possible to maintain this stance for long after the SED began more concentrated efforts to assure ideological as well as material integration of the scientific and technical intelligentsia from 1958 on. Public attacks on the allegedly "idealistic" character of Gottschaldt's position began in connection with a coordinated effort to create a "Marxist" psychology in the G.D.R. in 1958.[68] Ultimately the Socialist Unity Party's effort to gain the benefit from high-level work by non-party scientists while gradually increasing party influence in their institutes failed in this case. Gottschaldt received an offer from the University of Göttingen in August 1961, just before the Berlin wall went up, and left illegally in February 1962. He remained loyal to his research programme and methods until his death in 1991. Spectacular as his case was, it was not completely unique.

Summary. From the standpoint of the scientists themselves, these are all stories of successful struggle and scientific achievement against a variety of difficulties. From a broader perspective, however, the outcomes appear more ironic, almost paradoxical. These scientists used the positions they had achieved through academic networking and dealmaking with the state to institutionalize problem choices and research methods they had developed long before 1945. The effect, intended or not, was often to block innovation. In genetics and psychology, new ideas and methods, often imported from the United States or the Soviet Union, did not gain the upper hand in Germany until a younger generation of scientists wrested control from their elders in the early 1960s. Thus, precisely the success leading scientists had in reconstructing their careers was a primary contributor, along with lack of resources and continuing international suspicion of Germans, to relative scientific backwardness in both German states in the 1950s.

CONCLUSION

The development of science in postwar Germany combined considerable continuity in personnel, both in the Western zones and also to a greater extent in the Soviet zone than previously thought, with a subtle mix of effortfully constructed continuity and reconstruction in content. Science in postwar Germany should not be mapped along a grid with only two nodal points, "denazification" and "restoration." Scientists proved to be capable of adapting themselves and their practices to varied political and institutional circumstances, so long as available resources, local autonomy and the prestige accorded to science appeared adequate. The process of adaptation, in turn, had feedback effects on both the organization and the content of research.

That scientists got the chance to act out such adaptations at all depended upon the fact that, within certain limits, Socialist Unity Party officials in the G.D.R., like their West German counterparts and Allied military and civilian policymakers after 1945, regarded them as they presented themselves — as human resources whose knowledge and skills could be used regardless of their earlier political views. Historians would be well advised not to accept such assumptions uncritically, but rather to inquire more specifically into the multiple circumstances that allow or induce scientists, and science, to participate in projects of power.

NOTES

1 See, among many other works: Ulfried Geuter, *The Professionalization of Psychology under Nazism*, tr. Richard Holmes (New York and Cambridge: Cambridge University Press, 1992); Ute Deichmann, *Biologen unter Hitler* (Frankfurt am Main: Campus, 1992); Kristie Macrakis, *Surviving the Swastika: Scientific Research under National Socialism* (New York: Oxford University Press, 1993); Mark Walker and Monika Renneberg (eds.), *Science, Technology and National Socialism* (Cambridge: Cambridge University Press, 1994). For an overview, see Alan Beyerchen, "What We Now Know about Nazism and Science,' *Social Research*, 59 (1992), 615–641.

2 Herbert Mehrtens speaks of "self-mobilization" in "Kollaborationsverhältnisse: Natur und Technikwissenschaften im NS-Staat und ihre Historie," in Christoph Meinel and Peter Voswinckel (eds.), *Medizin, Naturwissenschaft, Technik und Nationalsozialismus — Kontinuitäten und Diskontinuitäten* (Stuttgart, 1994), 13–32.

3 Hans Mommsen, "Nationalsozialismus als vorgetäuschte Modernisierung," in *Der Nationalsozialismus und die deutsche Gesellschaft* (Reinbek: Rowohlt, 1991), 405–427.

4 For a brief overview of denazification policies and results in all four occupation zones with documents, see Clemens Vollnhals (ed.), *Entnazifizierung. Politische Säuberung und Rehabilitierung in den vier Besatzungszonen 1945–1949* (Munich: Deutscher Taschenbuchverlag, 1991).

5 On the situation in the new German states, see: Geoffrey J. Giles, "University History in the New German States: The Condition of the Archives," *German History*, 10 (1992), 366–391; Ralph Jessen, "Die Akten und das Chaos: Eine quellenkritische Zwischenbilanz zur Geschichte der DDR," *Deutschlandarchiv*, 25 (1992), 1202–1204.

6 For details, see the other contributions to this volume. See also: Ulrich Albrecht, Andreas Heinemann-Gruder and Arend Wellmann, *Die*

Spezialisten: Deutsche Naturwissenschaftler und Techniker in der Sowjetunion nach 1945 (Berlin: Dietz-Verlag, 1992); Macrakis, *Surviving the Swastika* (cit. n. 1), Epilogue; Hans-Uwe Feige, "Vor dem Abzug: Brain Drain. Die Zwangsevakuierung von Angehörigen der Universität Leipzig durch die US Army und ihre Folgen," *Deutschlandarchiv*, 24 (1991), 1302–1313.

7 See, e.g., John Gimbel, *Science, Technology and Reparations: Exploitation and Plunder in Postwar Germany* (Stanford: Stanford University Press, 1990), pp. 50, 69f.

8 "The Progress of Reeducation in Germany," Office of Intelligence Research Report 4237, 3 June 1947, trans in Alfons Söllner (ed.), *Zur Archäologie der Demokratie in Deutschland*, vol. 2 (Frankfurt a.M., 1986), esp. pp. 178f.

9 James F. Tent, "Denazification of Higher Education in US Occupied Germany, 1945–1949," in Manfred Heinemann (ed.), *Hochschuloffiziere und Wiederaufbau des Hochschulwesens in Westdeutschland 1945–1952* (Hildesheim: Lax, 1990), Teil 2, p. 12; Geoffrey J. Giles, "Reeducation at Heidelberg University," in Jürgen Heideking and Jürgen Herbst (eds.), *Mutual Influences on Education: Germany and the United States in the Twentieth Century* (in press).

10 Notker Hammerstein, *Die Johann Wolfgang Goethe-Universität Frankfurt am Main*, vol. 1 (Neuwied/Frankfurt: Metzner Verlag, 1989), pp. 590–591, writes that a volunteer "self cleansing" committee presented a list of 104 university teachers (from a total of 398) in Frankfurt to be removed as early as September 1945. Shepard Stone, in "Report on the Morale of Germany," *The New York Times*, 26 January 1947, wrote that 153 of 272 faculty members in Heidelberg had been released. For Erlangen, Munich and Marburg, see "The Progress of Reeducation in Germany," p. 185.

11 Tent, "Denazification of Higher Education," p. 12.

12 Lutz Niethammer, *Die Mitläuferfabrik. Entnazifizierung am Beispiel Bayerns* (Frankfurt a.M., 1982).

13 Manfred Heinemann, "Der Wiederaufbau der Kaiser-Wilhelm-Gesellschaft und die Neugründungen der Max-Planck-Gesellschaft (1945–1949)," in Rudolf Vierhaus and Bernhard vom Brocke (eds.), *Forschung im Spannungsfeld von Politik und Wissenschaft: Geschichte und Struktur der Kaiser-Wilhelm-/Max-Planck-Gesellschaft* (Stuttgart: Deutsche Verlags-Anstalt, 1990), esp. pp. 419f.

14 Heinemann, *Hochschuloffiziere* (cit. n. 9). Teil 1, p. 140.

15 David Philipps, *German Universities after the Surrender: British Occupation Policy and the Control of Higher Education* (Oxford: Oxford University Press, 1983); Ian D. Turner, "Denazification in the British Zone," in idem. (ed.), *Reconstruction in Post-War Germany: British Occupation Policy and the Western Zone, 1945–55* (Oxford: Berg, 1989), 239–267; Alan Beyerchen, "German Scientists and Research Institutions in Allied Occupation Policy," *History of Education Quarterly*, (1982), pp. 298–299.

16 Ulrich Schneider, "Zur Entnazifizierung der Hochschullehrer in Niedersachsen 1945–1949," *Niedersächsisches Jahrbuch für Landesgeschichte*, 61 (1989), 325–346, here: pp. 337–340.

17 Commission des Constituantes, report of 8 April 1946, cit. in Heinemann, *Hochschuloffiziere*, Teil 3, p. 41.

18 Heinemann, "Wiederaufbau" (cit. n. 13), pp. 417, 421.

19 Heinemann, *Hochschuloffiziere*, Teil 3, p. 238.

20 *Digest for Germany and Austria*, 2 February 1947, p. 7, cit. in "The Progress of Reeducation" (cit. n. 8), p. 187.

21 Hans-Uwe Feige, "Aspekte der Hochschulpolitik der Sowjetischen Militäradministration in Deutschland (1945–1948)," *Deutschlandarchiv*, 25 (1992), 1169–1180.

22 Otto Winzer (head of the Department of Education in the Berlin Magistrat), personal note to DVV President Paul Wandel, 21 September 1945, Bundes-archiv Abteilung Potsdam (= BAP) R-2 1307 B1. 27; see also BAP R-2 959 B1. 11f.

23 For an example of DVV frustration in this regard, see H.-J. Peters to Wandel 2 July 1947, BAP R-2 1438 B1. 63–64.

24 Alexandr Haritonow, "Entnazifizierung an der Bergakademie Freiberg 1945–1948," *Bildung und Erziehung*, 45 (1992), 434–436.

25 BAP R-2 1307 B1. 15–22 esp. 17f. See also Helga A. Welsh, "Entnazifizierung und Wiedereröffnung der Universität Leipzig 1945–1946. Ein Bericht des damaligen Rektors Bernhard Schweitzer," *Vierteljahreshefte für Zeitgeschichte*, 33 (1985), 339–372; Hans-Uwe Feige, "Zur Entnazifizierung des Lehrkörpers an der Universität Leipzig," *Zeitschrift für Geschichtswissenschaft*, 42 (1994), 795–808.

26 Feige, "Aspekte der Hochschulpolitik" (cit. n. 21), p. 1174.

27 So stated by Wandel in R-2 1307 (B1. 20), and explicitly reconfirmed in response to an inquiry from Saxony in October 1946, BAP R-2 958 B1. 129–131.

28 BAP R-2 955. Unfortunately, only (incomplete) lists of names are in this file, not the records themselves.

29 BAP R-2 943 B1. 23, 26f.

30 For Leipzig, Berlin, Greifswald and Dresden, see Feige, "Aspekte der Hochschulpolitik," p. 1174; for Jena and Rostock, and different figures for the other schools (e.g., 84.5 per cent for Berlin), see Roland Köhler, "Zur antifaschistisch-demokratischen Reform des Hochschulwesens der DDR 1945–1950" (Diss. A), Leipzig, 1969, p. 94.

31 BAP R-2 911, B1. 104. Unfortunately, the document gives no breakdown by discipline or university.

32 Ralph Jessen, "Professoren im Sozialismus. Aspekte des Strukturwandels der Hochschullehrerschaft in der Ulbricht-Ära," in Hartmut Kaelble, *et al.* (eds.), *Sozialgeschichte der DDR* (Stuttgart, 1994), pp. 226, 241. Unfortunately, the document cited by Jessen does not state how many of the SED members had also belonged to the NSDAP.

33 Michael Hubenstorf and Peter Th. Walter, "Politische Bedingungen und allgemeine Veränderungen des Berliner Wissenschaftsbetriebes 1920 bis 1950," in Wolfram Fischer, *et al.* (eds.), *Exodus von Wissenschaften aus Berlin* (Berlin: De Gruyter, 1994), esp. pp. 67ff.

34 Stiftverband der deutschen Wissenschaft, *Forschung heißt Arbeit und Brot* (Stuttgart: Steingrüben Verlag, 1950), pp. 17, 20. Despite its obviously tendentious purpose, this remains the most extensive available survey.

35 Burghard Ciesla, "Das 'Project Paperclip' — deutsche Naturwissenschaftler und Techniker in den USA (1946–1952)," in Jürgen Kocka (ed.), *Historische DDR-Forschung* (Berlin: Akademie-Verlag, 1993), p. 294; Albrecht, *et al.*, *Die Spezialisten* (cit. n. 6), p. 178.

36 Havemann to Shulits, copy in Hessisches Hauptstaatsarchiv Wiesbaden (= HSW), OMGUS/OMGH 8/22–2/8; *Neue Zeitung*, 3 May 1946; "Handlanger des Verbrechens," *Tägliche Rundschau*, 23 May 1946. For the following, see also Peter Weingart, et al, *Rasse, Blut und Gene*, pp. 572ff.

37 Verschuer to Hartshorne, 25 July 1946; Lt. Col. Charles Stewart (E and RA) to Minister of Education and Religious Affairs Land Greater Hesse, 13 August 1946; statement by Elsa Fay Hartshore, 24 October 1946. All HSW, OMGUS/OMGH 8/22–2/8.

38 HSW, 520/F (Ministerium für politische Befreiung — öffentl. Kläger), Nr. FZ 5261.

39 Stadtarchiv Frankfurt, Stadtskanzlei (Magistrat) 6603/18.

40 HSW 520/F, Nr. F2 5261. The formal disposition of the case is not stated in this file. Heimann memo, 14 May 1946, HSW OMGUS/OMGH 8/22–2/8.

41 See, e.g., Verschuer, *Wirksame Faktoren im Leben des Menschen. Beobachtungen an ein- und zweieiigen Zwillingen durch 25 Jahre* (Wiesbaden: Franz Steiner, 1954); Gerhard Koch, *Humangenetik und Neuro-Psychiatrie in meiner Zeit (1932–1978). Jahre der Entscheidung* (Erlangen and Jena: Palm u. Enke, 1993), esp. pp. 286, 292ff. For useful background, see Paul Weindling, "The Survival of Eugenics in 20th-Century Germany," *American Journal of Human Genetics*, 52 (1993), 643–649.

42 David C. Rife, review of Verschuer, *Wirksame Faktoren, American Journal of Human Genetics*, 7 (1955), pp. 83–84.

43 Verschuer, "Strahlenschädigung der Erbanlagen und Mutationsrate des Menschen," *Fortschritte der Medizin*, 75 (1957), 717–733; idem., "Das neue Institut für Humangenetik der Universität Münster/Westf.," *Homo*, 12 (1961), esp. pp. 5–6; Weingart *et al.*, *Rasse Blut, Gene*, pp. 589ff.

44 Jonathan Harwood, *Styles of Scientific Thought: The German Genetics Community 1900–1933* (Chicago: University of Chicago Press, 1993), pp. 204, 223–224; Macrakis, *Surviving the Swastika* (cit. n. 1), pp. 122f.

45 Harwood, *Styles*, p. 260 n. 113; Hans Stubbe, "Arbeit für eine gute Zukunft," in "*... einer neuen Zeit Beginn": Erinnerungen an die Anfänge unserer Kulturrevolution, 1945–1949* (Berlin and Weimar: Aufbau Verlag, 1980), 475–479.

46 Bericht zur Lage am Institut für Kulturpflanzenforschung in Gatersleben, 13 June 1955, Stiftung Archiv der Parteien und Massenorganisationen der DDR im Bundesarchiv Berlin (= ZPA (Parteiarchiv)), IV 2/904/422, B1. 361–373.

47 ZPA IV 2/904/422, B1. 350, 360.

48 Ibid., B1. 373.

49 ZPA IV 2/904/422, B1. 349, 378ff. The Institute's budget had been 1,773 million Marks in 1949, but had been radically cut in 1951 and 1952 to 844,000 and 900,000 Marks, respectively, probably due to the genetics departments' exclusion from the first five-year plan because of the Lysenko affair; by 1956 the budget was 4,386,600 marks.

50 "Bericht der Forschung der naturwissenschaftlichen, technischen und medizinischen Institute der Deutschen Akademie der Wissenschaften zu Berlin," 7 December 1957, ZPA IV 2/904/Band 415, B1. 38–39.

51 REM K 323 (Personalakte Oswald Kroh), BAP R-2 976.

52 Oswald Kroh, *Entwicklungspychologie des Grundschulkindes*, 11.–12. ed. (Langensalza: Beyer, 1935), e.g. p. 25; idem, "Die Psychologie im Dienst der völkischen Erziehung," *Bericht über den 16, Kongreß der Deutschen Gesellschaft für Psychologie* ... 1938 (Jean: Fischer, 1939), p. 43.

53 Ulfried Geuter, *The Professionalization of Psychology in Nazi Germany*, (cit. n. 1).

54 Universitätsarchiv der Humboldt-Universität zu Berlin (= UAHUB), K776 Band 2 B1. 5, 24, 31, 38–39; Band 3 B1. 19, 31, 32, 33, 39–50, 42.

55 BAP R-2 1067 B1. 1f.

56 Heise to Rompe, 16 July 1949 and "Fachliche Stellungnahme," 14 April 1948, BAP R-2 976 B1. 47, 56: Marquardt to Wandel 3 July 1948, *ibid.*, B1. 22.

57 Wandel, 10 June 1948, ibid., B1. 29; Rompe to Wandel 20 September 1948, *ibid.*, B1. 20.

58 Peter Mattes, "Zur Kontinuität in der deutschen Psychologie über die NS-Zeit hinaus," *Psychologie und Geschichte*, 1 (1989), 1–11; Alexandre Métraux, "Die Methodenstreit und die Amerikanisierung der Psychologie in der Bundesrepublik 1950–1970," in Mitchell G. Ash and Ulfried Geuter, *Geschichte der deutschen Psychologie* (Opladen: Westdeutscher Verlag, 1985), esp. pp. 236–244.

59 For the following see: Kurt Gottschaldt, interview with the author, Göttingen, 11 August 1987; idem., "Bericht über meinen wissenschaftlichen Werdegang," 1 April 1946, UAHUB G 395, copy in: Mike Stepputat, "Zur Geschichte der psychologischen Erbe-Umwelt-Forschung in Berlin 1935 bis 1942," Diplom. arbeit HU Berlin, 1990, unpublished.

60 Mitchell G. Ash, "Die Erbpsychologische Abteilung des Kaiser-Wilhelm–Instituts für Anthropologie, menschliche Erblehre und Eugenik 1935–1945," in Wolfgang Schönpflug and Lothar Sprung (eds.), *Geschichte der Psychologie in Berlin* (Frankfurt a.M.: Peter Lang, 1992), 205–222.

61 HSW, OMGUS/OMGH 8/22–2/8. As part of Verschuer's denazification proceedings, already described, Greater Hesse Ministerial Director Willy Viehweg sent Shulits' successor at E and RA in Berlin, Fritz Karsen, on 7 November 1946 a list of items transported from Solz to Frankfurt, including private property of Gottschaldt's as well as data from his twin research.

62 E.g.: Verschuer to Otto Hahn, 23 May 1946, 30 October 1946; Archiv der Max-Planck-Gesellschaft, Abt. II, Rep. 1A, X 97.

63 M. Wolfsohn to B.B. Ferencz (Office of Chief Counsel for War Crimes, Berlin Branch) 7 November 1946, p.3: "Als von Verschuer von Prof. Dr. Gottschaldt, dem Leiter der KWG (Institut für psychologische Forschung) (*sic*!) befragt wurde, ob er über die Einzelheiten in Auschwitz informiert sei, bejahte er dies." HSW HSW 520/F, Nr. F2 5261.

64 UAHUB G 395 B1. 17, 20–21.

65 UAHUB G 395 B1. 65.

66 Gottschaldt, Forschungsplan 1950, 12 January 1950, BAP R-2 1433 B1. 49; idem., "Zur Theorie der Persönlichkeit und ihrer Entwicklung," *Zeitschrift für Psychologie*, 157 (1954), p.1.

67 Gottschaldt, "Zur Theorie der Persönlichkeit," p. 2–3; p.4, n. 2.

68 Hans Hiebsch, "Aufgaben und Situation der Pädagogischen Psychologie der DDR," *Pädagogik* (1958), Heft 4; Ulrich Ihlefeld, "Zur ideologischen Situation der pädagogischen Psychologie," *Pädagogik* (1958), Heft 7. For the following see Mitchell G. Ash, "Übertragungsschwierigkeiten: Kurt Gottschaldt und die Psychologie in der Sowjetischen Besatzungszone und in der Deutschen Demokratischen Republik," in Siegfried Jaeger, *et al.* (eds.), *Psychologie im soziokulturellen Wandel — Kontinuitäten und Diskontinuitäten* (Frankfurt a.M.: Peter Lang, 1995), 286–294.

CHAPTER 6

Assessing the Damages: Forced Technology Transfer and the German Chemical Industry

Raymond G. Stokes

Germany's powerful chemical industry figured prominently in the German war effort through 1945; its postwar development would also help shape future German economic prospects. The industry therefore constituted a focal point of initial Allied policy and practice in postwar Germany. Government planning within each of the Allied countries, initial occupation directives to the Allied forces, the Potsdam Protocols, and subsequent Allied plans for German industry all specified a range of programs to curtail the chemical industry's war potential and to shape its future development. The complexity of the policymaking process was considerable: not only did Allied policy change through time, but each of the four occupying powers had different interests that they pursued in different ways. This, combined with the variety of programs undertaken by the Allies and the ambiguous nature of the programs' effects, makes it practically impossible to determine precisely the impact of forced transfer of technology on the German chemical industry.

Thus, instead of trying to develop precise estimates of the damages that forced technology transfer inflicted on German industry — or of the bounty that such transfers involved for firms in the former Allied countries, — this paper seeks to give some indication of the current state of our knowledge on the subject based on two major sets of sources: first, newly accessible evidence from industrial archives in what used to be the German Democratic Republic (GDR) and in the Federal Republic of Germany (FRG); and, second, a small but increasing number of scholarly studies. In addition, the paper sketches out some of the key issues and questions that remained to be investigated.

It is organized into three major sections. The first deals with forced technology transfer in chemical industry of the Soviet Zone of Occupation, the later

GDR. The second deals with the western zones, the later FRG. Finally, I offer some tentative conclusions. Based on what we know to this point, I contend that the effects of forced technology transfer were ambiguous, that is that it is impossible to describe them as completely positive or negative, whether from the point of view of the occupying powers or from that of the occupied. Neither Allied hopes for nor German fears of the forced technology transfer programs were realized, although one of the programs' unintended effects was to promote integration of each of the postwar German states into the sphere of influence dominated by one of the two superpowers and, simultaneously, to reinforce the growing division of German technological culture.

THE SOVIET ZONE OF OCCUPATION AND THE GERMAN DEMOCRATIC REPUBLIC (GDR)

Central Germany, the area what would eventually become the GDR, featured some of the largest and most modern chemical plants in the world. They included a number of major facilities that had belonged to the giant chemical firm I.G. Farbenindustrie A.G., the most important of which were the Leuna-Werke in Merseburg and the Buna-Werke in Schkopau. Because of the special importance of the Leuna-Werke to the German war economy (it produced synthetic fuel as well as a number of other vital chemical products), a total of 23 heavy air attacks were directed at it, with about 10,000–12,000 bombs landing directly on production facilities. The result was a destruction rate of about 50 per cent.[1] Of the major facilities of I.G. Farben, only BASF complexes at Ludwigshafen and Oppau suffered anywhere near this extent of destruction directly from the war.[2] Even the Buna factory, for example, which was very close to Leuna, suffered very little damage owing to bombing.[3]

Because the military front in late spring 1945 (as Allied troops completed their defeat of the Third Reich) did not conform to the boundaries of the postwar zones of occupation, most factories in what would become the Soviet zone were at first occupied by American military government personnel. The Americans pulled back to their own zone of occupation at the end of June 1945 according to Allied agreement. When they left, however, "they took with them the leaders of the [Leuna] works who had remained in office to that point and several outstanding scientists, chemists, and engineers, 27 in all."[4] The experience of the Buna works was similar: 25 of its chemists and engineers accompanied the Americans as they withdrew to the west.[5] What is more, the Americans also took with them important documents.[6]

Thus, the initial impact of Allied technology transfer in the eastern zone was substantial and apparent: key personnel and documents were removed from factories that were often already severely damaged by the war itself. Such removals, it seemed clear, would severely affect the industry's ability to reconstruct itself. Still, it is necessary to qualify this statement somewhat. After all, although many of these persons were undoubtedly interviewed by technical investigators working for the Western Allies (for the purposes of compiling technical reports, which we shall discuss later), many of them stayed in what would become West Germany and aided in the reconstruction of the chemical industry there. What is more, "part of them came back to Leuna."[7] We clearly need to know still more about the American action and its effects, but it is equally clear that American forays into forced transfer of technology paled in comparison to those of the Soviets.

In late June and early July 1945, Soviet troops occupied the Leuna plant as well as other factories in their zone. They, more than the Americans, actively pursued resumption of production in a number of different areas, but, by fall 1945, they also began to dismantle and transport back to the Soviet Union individual machines and artifacts from the Leuna (and other) works. The most important dismantling action, however, was carried out beginning in March 1946. Leuna deployed 30,732 of its workers and 7,376 other plant personnel to assist 400 Soviet officers and 1,000 to 1,200 soldiers from the Red Army to remove 120,000 tons of machines and structural iron and steel from the works. Included in the long list of affected installations were eight working compressors for synthetic gas, large-scale installations for methanol synthesis, and various machines, apparatus, and installations for synthetic gasoline production. What is more, the Soviets seized 117 journals and 514 books from the works library, in all a total of 1,067 individual volumes.[8]

While this list is impressive and makes it look as if it may have been disastrous for the Leuna Werke, there are good reasons to qualify this tentative conclusion as well. For instance, Soviet policy changed in mid-1946 from removals of physical plant to seizure of actual property of key firms in the Soviet zone of occupation (for more effective exploitation for the Soviet economy). The policy change entailed a number of other changes, among them the return of some of the artifacts that had been seized. Four of the eight synthesis gas compressors that had been removed, for instance, were subsequently returned "so that the production targets that the Soviet military administration had set in the meantime could be fulfilled." In addition, facilities for ammonia synthesis were also returned.[9] How typical were these examples? We do not know. But even if they were exceptions, the effects of dismantling remain unclear: certainly in the long run, it may have been more beneficial for the factory to lose older machines and apparatus if they were replaced soon enough by more modern ones.

There are a number of other issues linked to forced technology transfers from the Soviet zone of occupation that require considerable additional research. For instance, what were the similarities and differences between Soviet efforts to investigate and seize technical information and those of the western Allies? What about parallels between Soviet and western seizures of scientific personnel? In the west, it was unusual for personnel from the chemical industry to be seized and taken away to Allied countries, but not so in the Soviet zone. On 22 and 23 October 1946, Soviet troops appeared in Merseburg and the surrounding area and took away thirty-four chemists, engineers, and technicians. Among them were a number of highly qualified researchers, including Dr. Geiseler, who upon his return to the Leuna-Werke in the early fifties began the East German research program on high-pressure production of polyethylene. Geiseler and the others returned to East Germany between April 1951 and the end of 1952, but the precise negative impacts of their loss to the East German chemical industry for five years — and the probable positive impacts of their gain to the Soviet chemical industry during the same period — are not clear.[10]

This leads to yet another area in need of research. What was the impact of technology and personnel seized from the Soviet zone and East Germany on the Soviet economy? While we know little about what the chemists and engineers seized in autumn 1946 did during their five-year stint in the Soviet Union, there exists generally accepted wisdom about at least some of the technology. Reporting on opinions about dismantling at the Leuna Werke in 1946, P. Litsch noted that:

> The attitude of the work force toward dismantling was bad. There were opinions such as: the Russians are carting everything away, and: They can't use the stuff anyhow, it just sits around in train stations and rots.[11]

Was this — widely held — opinion accurate? If not, how useful was German chemical technology to Soviet industry?

Answering these questions is important, because it would shed additional light on one tentative conclusion that one may make about the results of forced technology transfer: in part owing to transfers of equipment and personnel, eastern German firms began to reorient themselves toward new suppliers of raw materials and semi-finished goods, toward new partners in technological development, and toward new markets for their products. In other words, forced transfers probably promoted integration of the GDR into the Soviet political and economic sphere.

Integration into the Soviet sphere did not have a positive impact on the GDR economically or technologically, and the GDR's technology — and in particular its chemical technology — began lagging considerably behind the west already during the 1950s, a lag that grew ever more serious in the following decades.[12] Explaining exactly what factors account for this lag constitutes a key task for historians, but I would suggest on the basis of the evidence available thus far that forced technology transfer was just one — and probably not a very important one — of a number of factors influencing this trend. Centralized state planning was one other major factor in the relative decline of East German industry, of course. But one might also add deficient oil reserves; lack of foreign exchange in a world economy increasingly dominated by western countries with hard-currency reserves; poor R & D work because of low budgets and poor equipment; emphasis on applications rather than basic research; and loss of key personnel to the west. Which of these caused poor performance? Which of them resulted from poor performance? One simply cannot say. Over time, however, these factors interacted with and amplified one another to assure that East Germany fell well behind the west in developing its industry and technology.

THE WESTERN ZONES OF OCCUPATION AND THE FEDERAL REPUBLIC OF GERMANY (FRG)[13]

Just as in the Soviet zone, physical removal of industrial plant and equipment took place in the western zones and was one aspect of technology transfer from Germany to its former enemies. Chemical plants represented a key policy objective. As the occupation began, well over half of the industry's facilities was slated for removal; even the revised level of industry plan issued in August 1947 for the combined US–UK zones stated that "between 40 and 50% of the total chemical capacity ... will ... be removed as reparations or destroyed.[14]

Still, this figure overstated the extent of dismantling: over three-fourths of the total capacity tagged for removal represented war-related explosives plants; in any case, actual dismantling never reached even these revised targets. A more reasonable estimate compiled in late 1949 pegged loss of postwar capacity for the chemical industry in the west at six per cent.[15] But even this figure is somewhat suspect owing to the broad range of firms and countries involved in the dismantling effort and their differing methods of valuation of seized assets.[16]

Variations in Allied dismantling policy and practice determined the differing experiences of chemical factories. American occupiers, finding little equipment worth carting away and charged with seeing that the German economy did not become a major burden for the US taxpayer, chose other means of realizing policy objectives: information, not hardware, was the preferred means of transfer. The Russians found much equipment that seemed worth removing, and the French and British ranked themselves between the two superpowers.

Given such variability, how is one to sum up the experience of western German chemical firms with regard to dismantling? For one thing, for the factories involved, losing some plant represented an unpleasant experience, but also had a positive aspect; it provided some impetus for technological improvement. On the other hand, one must not ascribe too much weight to this point. Few factories in the west lost more than a small fraction of their capacity. Most used prewar and wartime equipment to re-enter domestic and export markets.[17]

Despite dismantling, then, the overwhelming majority of capacity remained intact in western Germany. And, as Allied policies changed, all — including the latest — equipment fell from dismantling lists. Forced transfer of equipment through dismantling and reparations programs, therefore, was not a major factor impeding reconstruction in the FRG. Instead, inadequate transportation, fuel, and raw materials were the most important bottlenecks in industrial reconstruction and renewal.[18] The relatively low level of removal or transfer of plant and equipment meant that virtually all chemical production capacity that remained in the FRG area was based on coal-chemical technology; it and new capacity built after 1945 generally retained this coal basis well into the fifties.[19]

Allied efforts to seize "*geistige Reparationen*," or intellectual reparations, from the defeated country were much more important in the western zones of occupation than removals of plant and machinery. Allied teams gathered information on German wartime scientific and technological developments by investigating plant and interrogating personnel. Their primary purpose was to meet the military, industrial, and commercial needs of the former Allied countries. Information transfer also occurred when German scientists and technicians emigrated to Allied countries, although this form of technology transfer was apparently relatively unimportant for the chemical industry.[20]

Chemical technology constituted a primary target of Allied investigations, as indicated by an analysis of the lists of reports produced by Allied technical investigation teams. These reports — which were published and widely disseminated to college and university libraries as well as to industrial firms in the Allied countries — described the state of German technology in specialized areas. Chemical technology and, in particular, developments at I.G. Farben factories and research institutes during the war, figured prominently among them, and the published reports concentrated heavily on synthetic fuels, synthetic rubber, and acetylene chemistry.[21]

What the investigations brought to Allied firms is unclear. Some firms used the reports as aids in designing equipment or processes.[22] But the written documents were often not very useful on their own and may have actually promoted the recovery of German chemical firms by providing an incentive for Allied firms to turn to them for the "know-how" to make the written information usable.[23] This accounts in part for the apparently contradictory behavior of representatives of US firms: they often wrote to the Office of Technical Services and the Department of Commerce to comment on the positive impact of written technical reports, but also later denied the use or value of such information.[34]

In any case, there are other complex — and often ambiguous — issues that arise when one attempts to assess the effects of forced transfer on companies in western Allied nations. Information, which was the coin of the realm in western Allied seizures from Germany, is even more difficult to place a value on than machinery, yet it brings to the fore some of the same questions: was it ever used? How much and to what end? Did it allow companies in the Allied countries to avoid investing in research and development (R&D) or to bring a product to market faster than might otherwise have been possible? How much did the company save in R&D, marketing, and other costs by using information gleaned from Germany?

Answering these questions for specific industries, technologies, and scientific areas may well be possible, but it will be more useful still to shift the focus of investigation from Allied economic gain (or German loss) to the direct technological or scientific impact of the seized information or personnel. Although little scholarly work has been done in this area thus far, exploring the direct impact of the programs on the Allies and/or the Germans in specific scientific, social, or industrial contexts avoids the pitfalls of trying to estimate precise economic value while at the same time allowing the researcher to concentrate on assessing the significance of Allied intelligence-gathering efforts. Peter Morris from the Science Museum and Mark Clark of the University of Delaware are currently at work on a project on the transfer of acetylene chemistry and magnetic-tape technology from Germany to the United States after the war. The results of their research should demonstrate the value of focusing on specific cases of the positive and negative effects of forced technology transfer for both the Allies and the Germans.

An approach centering on specific instances of technology transfer would also be useful in trying to assess the impact of forced transfers on German firms, since evidence on their effects on German companies is also mixed. On the one hand, estimates of the value of Allied seizures of patents from German industry, which the technological investigation effort only enhanced, were high and appeared damning to German competitive prospects. In February 1947, for instance, a German patent attorney claimed that "American experts estimate the worth of confiscated German patents abroad to be several million dollars,"[25] while in 1952 the Munich-based economic think-tank Ifo-Institut pegged losses of patents and trademarks for the German chemical industry alone at $10 billion.[26]

On the other hand, as mentioned, printed technical information is useful only insofar as "know-how" is also available. Eventually, German patent-holders sold or bartered their know-how to minimize the effects of patent seizures. What is more, rapid technological change within the chemical industry meant that newly developed products and processes made up an increasing portion of a firm's revenue. In 1963, for example, Gilbert Burck wrote in *Fortune* that "today probably more than half the [European chemical] industry's revenues come from products that did not exist twenty-five years ago." For Bayer A.G. in particular, Burck went on, "nearly 60 per cent of ... world sales ... now come from products that did not exist fifteen years ago."[27] The same was true for other major West German chemical producers.

CONCLUDING REMARKS

For firms in the FRG, the unexpected payoff of the forced-technology transfer programs was to promote their economic and technological integration into the western world economy. Despite the motivations that lay behind them, technical investigations actually helped forge (or reforge) links between German and Allied scientists, technicians, and industrialists. Western Allied investigators came as conquerors to defeated Germany, but, in the course of doing their work, many of them established or reestablished professional contacts with their German counterparts. Companies in many countries were able to put German scientific and technical information to use in developing new products or processes. But they were often forced eventually, in part through the process of seizing the information in the first place, to return to the Germans for further information and know-how. When they did so, West German businessmen could function more as partners in the international business and technological community than as defeated enemies.

The impact of Soviet forced technology transfer on the East German chemical industry was undoubtedly more severe than that in the West, being both more extensive and consisting of large amounts of equipment as well as personnel and information. Yet, undoubtedly, there were also similar effects to those of the western effort that tended to promote economic, political, and technological integration within the eastern bloc. While this clearly did not necessarily redound to the benefit of East Germany in economic and technological terms, it had the pronounced effect of reorienting East Germany technologically, economically, and politically away from its traditional roots in the west and toward the east.

All in all, then, forced technology transfer programs undertaken by the former Allies after World War II were neither as useful as the victors hoped nor as devastating as the vanquished feared. Both the Allies and the Germans conceived of technology as disembodied hardware and information that could be physically removed from one place (to its permanent loss) and carted off to another place (to its permanent gain). Had they conceived of technology differently — as a process rather than a product, as dynamic rather than static, as social rather than disembodied —, they may have anticipated one of the unintended long-term legacies of such programs, which was to promote integration of each of the postwar German economies into a political-economic system dominated by one of the two superpowers. The process of integration in turn presupposed an additional important effect of forced transfers on the German chemical industry: the programs helped promote the division of Germany not only politically, but also in terms of its technological culture.

NOTES

1 VEB Leuna-Werke "Walter Ulbricht" [hereafter Leuna-Werke], Büro des Generaldirektors, "Die Bedeutung der chemischen Industrie für die Volkswirtschaft in der DDR 1945–1956," n.d. [ca. late 1956], Leuna-Werke A.G. Werksarchiv, Merseburg, [hereafter LWA] 3225.

2 Raymond G. Stokes, *Divide and Prosper: The Heirs of I.G. Farben Under Allied Authority* (Berkeley: University of California Press, 1988) p. 90.

3 "Zehn Jahre Aufbau — Zehn Jahre Friedensarbeit im VEB Chemische Werke Buna" (mimeographed paper, 1956), pp. 13–14, seen in Buna-Werke Betriebsarchiv, Schkopau.

4 Leuna-Werke, "Die Bedeutung der chemischen Industrie für die Volkswritschaft in der DDR 1945–1956;" John Gimbel, *Science, Technology, and Reparations: Exploitation and Plunder in Postwar*

Germany, (Stanford: Stanford University Press, 1990) pp. 16-17. An official history of the Leuna Werke claims that the Americans were most interested in those responsible for "lubricants production, the organic section, the hydrogenation section, the technical section, and the patent bureau." See Kreisleitung der SED des VEB Leuna-Werke, 'Walter Ulbricht', ed. *Geschichte der Leuna-Werke, 'Walter Ulbricht': 1945 bis 1981* (Leipzig: Dt. Verl. für Grundstoffindustrie, 1986) p. 29.

5 Kreisleitung der SED, Kommission zur Geschichte der Arbeiterbewegung, ed., *Zahlen, Fakten, Ereignisse. Chronik zur Geschichte der Arbeiterbewegung des Kombinates VEB Chemische Werke Buna* (n.p., n.d. [ca. 1975]), [no page number].

6 In an interesting sidelight, it may well be that one of those documents arrived back at Leuna via a circuitous route. Dr. Geiseler, head of the project started at Leuna in late 1951 to develop a pilot plant for production of polyethylene, and his staff relied heavily on previous research and development at the I.G. Geiseler's initial report on his research, as well as a more detailed memo on polyethylene produced by the organic chemicals section at Leuna, mention that the Leuna team built their apparatus based on the report of an extensive discussion of polyethylene technology at the 26th meeting of I.G. Farben's "Kuko" (Kunststoff-Kommision, or Plastics Committee) on 15 February 1943. Since Geiseler did not have access to documentation from I.G. files (which had for the most part disappeared), he used an English translation of the report printed in *Modern Plastics* magazine in 1946. The English translation, which resulted from Allied missions to investigate and exploit German wartime technological developments, thus served not only to inform British and American industrialists and engineers of the state of German technology, but also as a vehicle that aided the reconstruction of research in postwar East Germany. Dr. Geiseler, "Jahresbericht über die Versuchsarbeiten des Themas 58, Entwicklung einer Versuchsanlage zur Herstellung von Polyäthylen," n.d. [probably early 1954], LWA Bericht 1874; Organische Abteilung, "Aktenvermerk Betr.: Polyäthylen," 5 January 1953, LWA 9109.

7 *Geschichte der Leuna-Werke 'Walter Ulbricht'*, p. 19.

8 Dokumentation zur Chronik der wissenschaftlich-technischen Entwicklung der Leuna-Werke von 1945 bis 1961, (Leuna, 1982), [*Zahlen und Fakten zur Betriebsgesichte*, Heft 16] pp. 4, 7–9. Seen in LWA.

9 Quote from *Dokumentation zur Chronik der wissenschaftlich-technischen Entwicklung der Leuna-Werke von 1945 bis 1961, p. 8*. Information on ammonia synthesis from *Geschichte der Leuna-Werke 'Walter Ulbricht'*, p. 32. More generally, see Martin McCauley, *The German Democratic Republic since 1945* (NY: St. Martin's, 1983), p. 25.

10 This is based primarily on documents in LWA 1195; for complaints on salary and other issues by returning scientists and technicians, LWA 1170; for Dr. Geiseler's work, for instance, Rat zur technischen Entwicklung des

Werkes, "Bericht über die wissenschaftliche Forschung im Jahre 1953," 30 December 1953, LWA 6056.

11 Quoted in *Geschichte der Leuna-Werke 'Walter Ulbricht'*, p. 29.

12 See, for instance, Raymond Bentley, *Research and Technology in the former German Democratic Republic* (Boulder, CO: Westview, 1992), esp. chapter two.

13 Much of the following draws upon material in Raymond G. Stokes, *Opting for Oil: The Political Economy of Technological Change in the West German Chemical Industry, 1945–1961* (New York: Cambridge University Press, 1994).

14 "Revised Plan for Level of Industry in the Combined US–UK Zones of Germany: Joint Statement by State and War Departments," August 29, 1947, in US Department of State, *Germany 1947–1949: The Story in Documents* (Washington, D.C.: US Government Printing Office, 1950) pp. 361–2.

15 Bremer Ausschuss für Wirtschaftsforschung, "Gesamtumfang der Demontagen," [end of November, 1949], B1. 22, in Bundesarchiv, Koblenz, Z35/317. "Postwar capacity" is all intact capacity in 1945 plus machinery repaired or rehabilitated as of 1947/48 (about an additional 10% of 1945 capacity). Postwar western German chemical and chemical-technical capacity was approximately that of the equivalent area in 1936. "Demontagen in Westdeutschland," *Chemie-Ingenieur-Technik* (1951) *23*: 593–4, provides a critical review of the report.

16 Critiquing estimates of dismantling and reparations, Werner Abelshauser states bluntly that "an exact reckoning of the actual extent of reparations and their capacity-reducing effect on the fixed assets of West German industry is not possible." See "Neuanfang oder Wiederaufbau? Zu den wirtschaftlichen und sozialen Ausgangsbedingungen der westlichen Industrie nach dem Zweiten Weltkrieg," *Technikgeschichte* (1986) *53*: 264–6, quote on page 265. For another, more recent attempt to estimate the value of both physical and "intellectual" reparations, see Christoph Buchheim, *Die Wiedereingliederung Westdeutschlands in die Weltwirtschaft 1945-1958* (Munich: Oldenbourg, 1990) pp. 89ff.

17 Raymond G. Stokes, "Technology and the West German *Wirtschaftswunder,*" *Technology and Culture* (1991) 32: 1–22.

18 Werner Abelshauser, *Wirtschaftsgeschichte der Bundesrepublik Deutschland 1945–1980* (Frankfurt: Suhrkamp, 1983) p. 38, for zonal figures; Stokes, *Divide and Prosper*, pp. 95–6, for BASF performance; ibid., pp. 109–51, for factors impeding recovery of the chemical industry immediately after the war.

19 For more on this, see Stokes, *Opting for Oil*.

20 Allied organizations conducting the "exploitation" effort (as internal documents termed it) included the British Intelligence Objectives Subcommittee (BIOS), the Field Information Agency, Technical (FIAT), and a number of others.

The best overview of these efforts in John Gimbel, *Science, Technology, and Reparations: Exploitation and Plunder in Postwar Germany*, (Stanford: Stanford University Press, 1990), while Carl Glatt of the European University Institute has a forthcoming dissertation on the British effort. My thanks to Mr. Glatt for allowing me to look at an early draft of a chapter on the Darwin Panel. An earlier, still useful, work on American use of German scientists in Clarence Lasby, *Project Paperclip: German Scientists and the Cold War* (New York: Athenaeum, 1971), esp. pp. 18–26, 129–130, and 163–4. On technical investigations, see e.g.: Leslie E. Simon, *German Research in World War II* (New York: John Wiley, 1947); Arnold Krammer "Technology Transfer as War Booty: The US Technical Oil Mission to Europe, 1945," *Technology and Culture* (1981) 22: 68–103; and Peter Spitz, *Petrochemicals: The Rise of an Industry* (New York: John Wiley, 1988), who deals specifically with chemical industry investigations, on pp. 2–17, 48–52, 113–14, 303–4, and 346.

21 I base my remarks on the subjects the Allies investigated on an evaluation of the "Alphabetical Subject Index of CIOS, BIOS, FIAT, and JIOA Final Reports," n.d., a copy of which is in Public Record Office, London (Kew), FO 1005 FO 1005/1602. John Gimbel describes some of the missions to study processes (such as Reppe chemistry and synthetic rubber) developed and employed at I.G. Farben facilities in *Science, Technology, and Reparations*, e.g. pp. 9–15, 30–31, 88.

22 Spitz, *Petrochemicals*, pp. 176, 329–31; Gimbel, *Science, Technology, and Reparations*, pp. 95–101, 224.

23 For instance, BASF L. -K. Abteilung, "Aktennotiz: *Betr.* Besuch der Herren von der Mitsubishi Chemical Industries Limited," July 27, 1953, BASF Unternehmensarchiv, Ludwigshafen, F9/79; Spitz, *Petrochemicals*, pp. 329–31. This experience is not surprising in light of scholarly research: see Keith Pavitt, "Technology Transfer among the Industrially Advanced Countries: An Overview," in Nathan Rosenberg and Claudio Frischtak, eds., *International Technology Transfer: Concepts, Methods and Comparisons* (New York: Praeger, 1985) pp. 3–23.

24 Gimbel: *Science, Technology, and Reparations*, pp. 96–101, 224, gives examples of these apparently contradictory comments. There were, of course, also compelling commercial and legal reasons for the firms to take this apparently contradictory tack.

25 Werner Colrausz, "Vorschläge für das Patentgesetz," *Angewandte Chemie* (1947) 19, Teil B: 50.

26 Ifo-Institut, *Chemie einschl. Kunststoffe* (Munich: Ifo-Institut, 1952) p. N5.

27 Gilbert Burck, "Chemicals: The Reluctant Competitors," *Fortune* (1963), *68*: on pp. 148 and 152.

CHAPTER 7

German High Velocity Aerodynamics and Their Significance for the US Air Force 1945–52

Burghard Ciesla

Compared to other American users, after 1945 the American Air Force[1] brought the highest number of Austrian and German scientists and technicians to the United States. The Air Force especially looked for experts who had been involved in Germany's previous research efforts in high-velocity aerodynamics.[2] The following paper deals with both the motives the Air Force had in hiring this particular group of German scientists and the results the application of German knowledge had in the US.

THE STATE OF GERMAN AND AMERICAN AERODYNAMIC RESEARCH AT THE END OF THE WAR

Various studies on the state of Germany's aerodynamic research efforts come to different conclusions. Some, like the one by R.J. Overy, conclude that Germany's efforts were barely effective and were inadequate.[3] Other analyses of German aircraft research show that, despite the disadvantages described by Overy, Germany developed some highly sophisticated airplanes and rockets. Already in mid-1944 the V1 Flying Bomb, shortly afterwards the V2 rocket, and the first series of Me 262 jet fighters, had shown that German aircraft research was in the lead in some fields.[4]

Even though new German aircraft weapons could not change the outcome of the war, they showed the strength of German's aircraft research in some important fields. Similar projects on the Allied side either did not exist at all or were still in their earliest stages.[5]

This point of view was even shared by some American military experts shortly after the end of the war. One example is a report of the Scientific Advisory Group of the US Army Air Force (AAF-SAG). Entitled *Where We Stand*, it was presented by the Group's chairman, Theodore von Kármán, in 1946. Based on investigations of the German aircraft industry conducted after the end of World War II, this report confirmed that American aircraft research had fallen behind.

One purpose of the report was to convince high-ranking people in policy and in the military about the need to strengthen American efforts in various fields of aerodynamic research.[6] In a clear effort to dramatize the situation, von Kármán indicated that the title of the report should have been "Where We Shall Go."[7] According to von Kármán, the United States would face difficulties in future if the country continued its aircraft and rocket development only on the basis of then-existing research facilities.

Fields where American research was behind that in Germany were above all rockets and guided missiles, jet aircraft and synthetic fuel.[8] Germany's lead was especially great in high-velocity aerodynamics.[9] This lead did not result from better engineering skills in Germany but from Germany's concrete experience in the testing and production of modern air technologies.[10]

That the Germans seemed to be more experienced in applied industrial research resulted from the conditions of war in Germany and not — as Overy has indicated — from the efforts of German scientists to look for a retreat or even forms of resistance against the Nazi regime. It is important to realize how intensively Germany as a power at war had been looking for new technological solutions in the area in which its failings were most obvious.

Germany's inferiority was especially great in the field of air war. In the German view, the massive Allied bomb attacks against targets in Germany and German-controlled areas would remain "unanswered" unless appropriate rockets and jet fighters were developed to respond to them. Therefore, one can say that Allied air attacks were the major reason for mobilizing every available resource in Germany for the development and production of rockets and other aircraft equipment.

Germany implemented a military-style guidance of its aircraft production that encouraged especially applied industrial research, for example in the field of high-velocity aerodynamics. The general military direction of Germany's aircraft production made it easier to put decisions into effect quickly. One has to add that both right and wrong decisions were carried through to completion under these circumstances.[11]

The promotion of high-velocity aerodynamics in Germany was also based (at least indirectly) on the effects of the Treaty of Versailles of June 28, 1919. According to the treaty Germany was not allowed to have an air industry of its own after World War I. On the one hand, Germany got around this prohibition by cooperating with the USSR.[12] On the other hand, the Versailles Treaty's regulations forced Germany into precisely that area of research that was to become most creative for developments in the aircraft and space industries.

Beginning in the early 1930s Germany established new research facilities that varied from, and were even not necessary for, traditional aircraft research.[13] One important example of new research facilities which could bypass the Versailles Treaty's restrictions was the construction of windtunnels. By the summer of 1944 German aerodynamic research had access to more than 78 windtunnels. Fifteen of them were located in occupied territory, for example in Prague, Vienna, Graz, Amsterdam and Warsaw.

Except for the case of the Vienna windtunnel, the most important research and development work remained confined to institutions within the traditional boundaries of Germany as of 1937. Here there were 63 windtunnels available. Fourteen of them were established in the framework of German technical universities,[14] 39 belonged to special air research centres, while 10 were owned by the German aircraft industries.[15]

The technical capability of each of the 78 windtunnels was different. Five windtunnels were still under construction, including the world's largest high-velocity windtunnel (40″ × 40″) in Kochel, capable of ten Mach.[16] Only 19 of all 78 windtunnels could be used for high-velocity aerodynamics and even only 13 were useful for high-velocity tests. The most important of the latter were located in Berlin, Kochel, Göttingen, Ötztal near Munich, Aachen and close to Braunschweig.

Most of the windtunnels were built during the Nazi era. Between 1933 and 1938, i.e., the period of rearmament, Germany built 29 windtunnels. Another 33 were constructed during the World War II.[17] The Allies found the most powerful windtunnels at Kochel and Ötztal. In addition, they conquered jet engine test centres like the one at the BMW Herbitus works in Munich. The Allies found that throughout Germany at least 14 important research centres had contributed to German aircraft and rocket research. Most of the facilities were concentrated in Peenemünde and at the Luftfahrtforschungsanstalt (Aircraft Research Institute) in Völkenrode, the latter established when the first institute had been evacuated to the Harz region during the war.[18]

Compared to this, the situation in the United States was quite different. US aircraft and rocket development had fallen behind that in Germany with backlogs in certain areas of between two and fifteen years.[19] Although the United States was provided with a significant research program in aerodynamic testing, its high-velocity research was equipped with only 3 windtunnels, including the one at Boeing that only became available in summer 1945,[20] though it had been planned before America's entry into the war in December 1941.[21] The other two windtunnels belonged to the National Advisory Committee for Aeronautics (NACA). In addition, the capability of the American windtunnels was limited compared to the two most modern German plants in Ötztal and Kochel.[22]

The weakness of American research is also apparent if we not only compare numbers but also consider the access which firms had to windtunnels in Germany and the US, respectively. Almost every large German aircraft company had its own high-velocity research installation or quick access to those of other research centres. On the other hand, during the development phase of the first of the Army Air Force jet bombers in 1944 aircraft companies had to "line up" at the NACA.

Over-demand severely limited the time available for testing in the windtunnels. It is no wonder that the jet prototypes of competing American companies looked very similar, since the very same NACA teams assisted or led development teams from Convair, Martin, Boeing and North American.[23]

THE TRANSFER OF GERMAN SCIENTIST AND TECHNOLOGY AND ITS EFFECTS

Operation Overcast (July 1945–March 1946) and Project Paperclip (beginning in March 1946) mark the postwar removal of German technology to the USA. In view of the backwardness of, for example, American windtunnel research, it is no wonder that US authorities quickly decided to make use of German-built windtunnels made available through American occupation of Germany. In addition, US authorities capitalized on the new "access" to the scientists and technicians who were able to use these windtunnels.

There is a lot of literature about Project Paperclip.[24] Otherwise, the role of the Army Air Force and its Scientific Advisory Group played in the transfer of technology from Germany is yet to be thoroughly analyzed.[25] The analysis shows that the major task of this group was to evaluate German air research projects and to select those that should be transferred to the USA. Project Paperclip, then, dealt with the transfer, the adaptation and the application of German research capabilities in the US itself.

The idea to establish special Army units that should evaluate captured or occupied German aircraft research facilities from Army General Henry H. Arnold. In late 1944 he initiated the foundation of the Scientific Advisory Group of the Army Air Force (AAF-SAG).[26] Arnold, who was aware of the gaps in American aircraft and rocket development, contacted Theodore von Kármán in September 1944.[27]

Arnold offered the position of AAF-SAG chairman to von Kármán after he had explained his plan to create this particular scientific advisory group for the Air Force. Arnold's idea was to bring together the leading American specialists from various areas of air research in one specific unit. Like the ALSOS program that had to secure the capture of German nuclear physicists and their research results, the AAF-SAG had the same aims in the field of aircraft research.[28] Von Kármán, seeing the opportunity for a significant promotion of American air research, agreed to bring German talent to the Air Force, as Arnold had suggested.[29]

Since von Kármán had studied, taught and researched in Göttingen and Aachen until the late twenties himself, he was familiar with Germany's previous working results in the field of aircraft research.[30] He had stayed in contact with his colleagues in Germany until the beginning of the war. These contacts resumed after the end of the war, when von Kármán made his first trip to Germany for the AAF-SAG in May 1945.[31]

The involvement of people who were familiar with the proceedings in Germany's scientific community had the major advantage that these experts could quickly understand which of Germany's projects were interesting enough to be continued in the US, for example in the framework of Air Force projects. In the following time period members of AAF-SAG were always the "starting point" for American air research projects that had previously been conducted in Germany.

Within a few weeks after the end of World War II in Europe a domestic American windtunnel program was established. F. L. Wattendorf, who had studied research facilities in Germany, came up with this idea after his return from Germany in June 1945.[32]

Wattendorf also suggested the establishment of Air Force research centres that could use research documentation confiscated in Germany. The capture of German research material had already gotten underway immediately after the end of the war under the guidance of the Air Force's Air Technical Intelligence.[33] With only a few exceptions it encountered no problems. The securing of the confiscations was its task rather than that of the AAF-SAG.

While technical documents were being confiscated in Germany the potential usefulness of German scientists became obvious. German specialists who had been involved in the writing of these documents were indispensable to the examination and evaluation of this material. Thus, the next stage of the process aimed at hiring these "specialists" and bringing them to the USA.

However, this part of the program was more difficult for the AAF-SAG. First, the American Air Force had to compete with other units of the United States' Armed Forces as well as with the other Allies and their armed forces. Second, right from the beginning Operation Overcast suffered from difficulties concerning the denazification of "candidates" and their entry into the USA.

For these reasons, the AAF-SAG's choice of most important scientists often remained wishful thinking. In May 1945 they had expected that preparations for the transfer of Germans would be completed quite rapidly.[34] However, it was only by March 1946, when Project Paperclip had come into operation, that these problems were solved on the American side.

Despite these initial problems, the efforts of the AAF-SAG to bring German air research scientists to the USA were quite successful. The US Air Force brought the greatest number of German scientists to America. Between fall 1945 and summer 1948 the Air Force brought about 200 Germans to the US with the help of Project Overcast/Paperclip. This figure remained almost constant until 1952.[35] Thus, the US Air Force was responsible for around forty per cent of the total 518 German scientists transferred to the USA by Project Paperclip between 1948 and 1952. The other branches of America's armed forces got 34 per cent (army) and 16 per cent (navy), respectively, whereas the Department of Commerce hired 10 per cent of the German scientists shipped to America.

Even though Project Paperclip became a permanent program in the 1950s, and was later replaced by the so-called "Project 63", the basic work of the program was done in its early phases. This is because about 90 per cent of the German "staff" hired between 1945 and 1952 stayed in the USA for good. Furthermore, initially hired Germans were replaced by others in only a few cases.[36]

SOME EFFECTS OF THE TRANSFER

Many major points of interest stand out in the investigation of the effects of technology transfer from Germany and the hiring of German scientists. Without pretending to cover them all, we focus the discussion on the following five:

1. the effects the technology transfer had on basic research;
2. the consequences it had for civilian personnel in the military;
3. the cooperation between German scientists and American industries;
4. the way information was gathered and processed and, finally,
5. the way American research was organized.

1. About 80 per cent of the German scientists employed by the US Air Force did their work at Wright Field, Ohio.[37] Although they represented a variety of research fields, about a quarter of them were specialists in aerodynamics.[38] Here they dealt mainly with high-velocity research using windtunnel technology.

The most successful working groups in windtunnel technology adapted to the interests of the US Air Force were the ones led by the German scientists Goethert and Hermann. Both had a great deal of experience with windtunnels. A report on Hermann's group reads as follows: "The assistance obtained from Dr. Hermann and his organization on the continuation of a German Mach number 10 windtunnel has saved this country at least five years of work: The utilization of this organization of scientists who have had fourteen years of experience in supersonic wind tunnel design and operation has saved and will continue to save the country huge amounts of time and money in this field."[39]

In addition, the Germans did theoretical and experimental research work on "Transonic Range", the "Interference Veiling Problem" and "Delta Wing Configuration."[40] In addition to aerodynamics the US Air Force put German engineers into jet and rocket research, avionics, air medical research and aircraft construction. A quarter of the German Air Force specialists[41] concentrated on a range of issues related to such fields as aerial photography, optics, fuel development, materials research, parachuting and meteorology.[42]

This research work was not kept isolated from the rest of US research. American authorities looked for the integration of the German scientists into the American scientific community. Thus, it is no wonder that the Germans were able to see their results going straight into current Air Force projects run by American scientists. In contrast to this German scientists who had been moved east, remained isolated from Soviet scientists in the same research fields.

2. The influx of German scientists solved other problems the US Air Force faced at the end of the war. After 1945 a number of highly qualified American scientists employed by the Air Force wished to return to their employers in the private sector. Unable to match the salaries private companies offered to these scientists, the Air Force experienced gaps in qualified research staff in a range of fields.

Project Paperclip provided the almost ideal solution to the problem. Scientists hired in Germany replaced Americans, who rejoined their previous employers. In addition, the Air Force military found that the Germans were easier to control than the Americans.[43] Moreover it could be taken for granted that political and ideological views were a lot less important to the Germans than their own particular research interests.

In a memorandum to the Director of the Joint Intelligence Objectives Agency (JIOA) of May 4, 1948 it was observed that "the Americans are very well impressed with the sincerity and cooperativeness of the Germans. No German has been found prying into things which were not his business. The Americans in charge feel more concern over the possibility of leaks through other Americans they employ than through the Germans. They point out that information in the hands of five hundred marked and suspect Germans is safer than that in the hands of a great number of Americans who have no finger pointing at them."[44]

Last, but not least, the Germans' work for the Air Force was advantageous in its economic results. The value of German research, for example, at Wright's Field was estimated by the Air Force in September 1947 to be thirty million dollars. Even though one has to be skeptical about this estimation, since Wright's Field's budget in 1948 was only two billion dollars, it shows how highly the Air Force appreciated the work done by the Germans.[45]

3. One of the most significant results the presence of German experts in the US had was the development of close relationships between the German civilian personnel in the military on the one hand and industrial companies on the other. By 1948 representatives of industrial companies had made about 500 interviews with German scientists at Wright's Field. Industry's interest remained undiminished after that date.[46]

There were many fields of research where the Germans were of value for private companies. They could help explain developments that had been taken over directly from German research facilities. They could also help deal with developmental problems cropping up in new projects, either as advisors or simply as partners in discussion. Last not least, the contacts created the opportunity for some of the Germans to prepare their subsequent move into industry when their contractual obligation with the Air Force had expired.

The contacts between the Germans in the military and private companies were of great value. For example, in March 1947 North American Aviation reported that the company had saved approximately 40,000 dollars in the construction of a high-velocity windtunnel by using advice given by the German scientist Dr. Hermann. General Electric also reported at the same time that it reckoned its discussions with another German, Dr. Doblhoff, would be worth a million dollars.[47]

4. The huge number of research reports and the captured equipment found in Germany's air research centres were something of a problem for Allied evaluation teams. To properly assess Germany's research and development capability in the air industry a systematic collection and processing of the documents was necessary. For that purpose two facilities were established.

With British approval both the Air Force and the Navy established the Air Document Research Centre in London.[48] In addition, the Air Document Research Office (ADRO) was founded at Wright's Field, Ohio. The employees of these centres had to collect, evaluate, catalogue, and film all available documents on Germany's air technology. Thereafter the materials were divided between the USA and Great Britain.

ADRO dealt with the efficient distribution of German documentation within the USA itself.[49] First, ADRO had to decide to what degree material could be passed on to potential American users.[50] After all, about 1,200 Air Force contractors, i.e. all of the companies that had contracts with the US Air Force, got access to information gathered at German air research facilities.

By May 1946 1200 tons of scientific and technical reports and 2500 tons of plant and equipment had arrived in London at Wright's Field.[51] The AAF was fully aware of the value of the captured documents. An AAF summary reads as follows: "For approximately 15 years the Germans had been preparing for World War II. Millions of dollars were spent in the construction of aeronautical research establishments, developments and production facilities. Millions more were spent for the research, engineering and production or work carried out at the facilities. If we are not too proud to make use of this German-born information, much benefit can be derived from it and we can advance from where German left off."[52]

The gathering, processing, and transmitting of German research results also indirectly stimulated the movement towards a modern information and library science, creating innovative methods of information processing.

5. Besides the documents per se that the US Air Force captured in Germany, the Americans could learn a lot about Germany's way of organizing research and development, including the logistics of research establishments. Parts of the German approach to organizing research were adapted to various research programmes run by the Air Force, NACA, or even the NASA in the following years. One example is the Unitary Windtunnel Plan of 1949, initiated by the AAF-SAG.[53] Another is the establishment of the Air Force's Arnold Engineering Development Center (AEDC) in 1951. In addition to experts from the American side like Wattendorf and von Kármán, the planning for this center even included the former director of the Peenemünde army base General Dornberger. During the planning stage for a future Air Force research institute all prewar and wartime experience from Germany's research centers was taken into consideration. The result was that the AEDC combined Peenemünde, Völkenrode and the Kochel and Ötztal windtunnels.[54]

CONCLUSION

With the help of air research documents captured in Germany and the hiring of German scientists, the US could make good on its wartime neglect of high-velocity aerodynamics. The American research efforts were guided by a different set of research priorities and war-related demands. Beginning in 1942 the Allies had concentrated their efforts on the build-up of strategic bombing units. This led to an approach which Senator Albert D. Thomas characterized after the war as follows: "We won the war with the Germans with brawn not brain. We choked them with the weight of our planes."[55] Indeed, the Manhattan Project to create an atomic bomb shows just where US priorities in armaments research lay.

Germany, by contrast, had to respond to the threat of Allied strategic bombing, and did so by promoting its own research efforts for the development of jet aircraft and rockets. Since both research projects had gotten under way already in the 1930s, it was no wonder that, at the end of the war, high-velocity aerodynamics in Germany were at a more advanced level than in Allied countries.

The two different approaches followed by Germany and the US during the war also show that technical innovation and basic research are assisted by different combinations of means and motivations. These circumstances can rapidly carry actors in wartime or crisis situations in different directions.[56] During the ALSOS mission the AAF-SAG's experts had expected to find considerable progress in Germany's atomic research. Instead, they found highly developed high-velocity aerodynamics.[57]

The impact of Germany's research and development efforts in the US was far greater in the way the Germans organized research centers. The establishment of the Air Force's AEDC followed by consistent the evaluation of both the way the Germans had done scientific research before 1945 and American programmes such as the Manhattan Project. In a study by Rolf Kreibich, recently published in Germany, the author emphasizes that the US Air Force got convinced shortly after the end of the war about the merits of the Germany way of research organization. The combination of research potential, resources and finance had secured the achievement of many important inventions in a very short time.[58]

The copying of German's way of organizing research was caused by pragmatism, too. When World War II ended, the time of limitless access to financial resources for the military was over. The employment of German specialists and the application of their skills helped overcome staff problems and financial restraints in the Air Force in this situation.

Moreover, with the employment of Germans at Air Force development centers American firms also had access to these scientists. The Junkers jet specialist Anselm Franz commended that "They sent me wherever something had gone wrong, so at least I learned how not to do things."[59] In the long run, these contacts meant for the Germans that they could keep their knowledge up-to-date and could even promote their future employment in private US companies.

When the Cold War began, German specialists in Allied countries indirectly helped re-evaluate the then West German scientific potential.[60] It is surprising that under these circumstances the American transfer of German technology at the end of World War II has received only scant attention. This is regrettable since an analysis of this transfer could contribute to explaining the priorities of postwar American foreign policy with regard to Germany and to Europe as a whole. In this regard some argue that US policy towards Germany in particular had no clear priorities in the early post-war years due to the fact that the politics of various parties were competitive with each other. However, it becomes obvious that these priorities actually existed, based on the traditional attitude of the American governing elite to shape politics by security concerns. As regards the emerging Cold War, American technology transfer from Germany can be seen in a different light. It shows how very quickly a new enemy, the Soviet Union, was identified and how the concept of containment became the leading motive of American foreign policy.[61]

Last but not least, a financial assessment of the transfer of scientists and technology from Germany to the US as a form of intellectual reparations does not seem to be a very worthwhile line of inquiry. The German economic historian Christoph Buchheim has very accurately written in an excursus on the reparations extracted by Western occupying powers that "An evaluation of the economics of reparations may provide some insight, but appears to be largely pointless from a political and legal point of view. Without doubt the occupying powers were the absolute masters and therefore in a position to impose obligations on Germany as they liked …, which Germany was then forced to fulfil. From the point of view of the political power to push through such demands the weighing of assistance on one hand with reparations on the other would be trying to balance things that cannot realistically be compared."[62]

NOTES

1 In 1947 the Army Air Force became one of the armed forces in its own right, from then on known as the United State Air Force (USAF).

2 See: National Archives and Record Administration (NARA), Record Group (RG) 18. *Memorandum for Assistant Chief of STAFF, Return of German Scientific Personnel for Temporary Technical Exploitation*, Headquarters, Army Air Force 9 July 1945, Box 12.

3 R.J. Overy, *The Air war 1939–1945* (USA: Scarborough House Publishers, 1991), 186–191.

4 Heinrich Klein, *Vom Geschoß zum Feuerpfeil, Der große Umbruch der Waffentechnik in Deutschland 1900–1970. Eine Dokumentation* (Neckargemünd: Kurt Vowinckel Verlag, 1977), 191–200; Adolf Galland, *Die ersten und die Letzten, Jagdflieger im Zweiten Weltkrieg* (München: Wilhelm Heyne Verlag, 1989), 196–304. Walter Dornberger, *Peenemünde,*

Die Geschichte der V-Waffen (Frankfurt/M-Berlin: Verlag Ullstein GmbH, 1989). Lothar Gruchmann, *Der Zweite Weltkrieg dtv-Weltgeschichte des 20, Jahrhunderts* (München: Deutscher Taschenbuch Verlag, 1990), 295–297. Ernst Stuhlinger/Frederick I. Ordway, *Wernher von Braun, Aufbruch in den Weltraum, Die Biographie* (Esslingen/München: Bechtle Verlag, 1992).

5 Gruchmann, "Der Zeite Weltkrieg", 296. Overy, "The Air War", 195. I. B. Holley (Jr.) "Jet Lag in the Army Air Corps", in Harry R. Borowski (ed.), *Military Planning in the Twentieth Century* (Washington D.C.: US Air Force Academy, 1986).

6 See: Foreword to: Kármán, "Where we stand", IV.

7 Ibid.

8 Bolling Air Force Base (BAFB), Historical Office, Mikrofilm (MR) A2055, *Speech given by Col. Putt at Dayton Club on 7 May 1946*, frame 0456.

9 BAFB, MF A2055, frame 0456, Kármán, "Where we stand", 1–47.

10 National Air and Space Museum (NASM), *Theodore von Kármán, U.S.A.F. Arnold Engineering Development Center, Brief History 1950*, Box 89.1/2.

11 Overy, "The Air War", 186–191. Klein, "Vom Geschoß zum Feuerpfeil", 85.

12 See: Heinrich Schwendmann, Die wirtschaftliche Zusammenarbeit zwischen dem Deutschen Reich und der Sowjetunion von 1939 bis 1941. Alternative zu Hitlers Ostprogramm? (Berlin: Akademie-Verlag, 1993); Olfa Gröhler, Selbstmörderische Allianz. Deutsch-Russische Militärbeziehungen 1920–1941 (Berlin: Vision 1992).

13 Dornberger, "Peenemünde", 28–33.

14 This means Technische Hochschulen.

15 According to Wright Patterson Air Force Base (WPAFB), Historical Office, *Technical Intelligence Supplement*, Headquarters Air Material Command, Dayton, Ohio 1946.

16 WPAFB, Technical Intelligence Supplement. Maxwell Air Force Base (MAFB), Historical Office, *Ötztal Wind Tunnel*, AAF Technical Report 5240, Headquarters Air Material Command, Dayton, Ohio 1945, Sig. 216.2101–5240. Kármán, "Where we stand", 1–8.

17 WPAFB, Technical Intelligence Supplement.

18 WPAFB, *Facilities Chart-German Aircraft 1946*, Air Technical Service US AAF 1946. WPAFB, *Production Chart-German Aircraft Equipment 1945*, Air Technical Service Command US AAF 1945. BAFB, MF A2055, frame 0452. Leslie E. Simon, *German research in World War II, An Analysis of the conduct of research* (New York: John Wiley & Sohns, Inc., 1947), 1–39. Kármán, "Where we stand".

19 BAFB, MF A2055, frame 0456.

20 National Air and Space Administration (NASA), *A Historical Study of the Benefits Derived from Application of Technical Advances to Civil Aviation*,

Vol. II, Washington DC 1971, Appendix C (Bellevue/Washington, State: TYC Publishing Company 1991), 138.

21 Cook, "The road to 707", 138. NASM, Theodore von Kármán Box 57.17 1/2 NASM, Theodore von Kármán, *Boeing Airplane Company 1934–1943* Box 57 Sec. III.

22 F. Zwicky, *Report on certain phases of war research in Germany* (Wright Field, Dayton, Ohio: Headquarters Air Material Comman, 1947), 3ff.

23 Cook, "The Road to 707", 138.

24 M. Bar-Zohar, The Hunt for the German Scientists (London: 1967). Clarence G. Lasby, Project Paperclip: German Scientists and the Cold War (New York: 1971). Frederick I. Ordway III/Mitchell R. Sharpe, *The Rocket Team* (New York, 1979). Kranx Kurowski, *Alliierte Jagd auf deutsche Wissenschaftler, Das Unternehmen Paperclip* (München: 1982). Linda Hunt, "US coverup of Nazi scientists", in *Bulletin of the Atomic Scientists* (1985) 4. Tom Bower, *The Paperclip Conspiracy, The Hunt for the Nazi Scientists* (Boston/Toronto: List Verlag, 1987). John Gimbel, "The American exploitation of German technical know-how after World War II", in *Political Science Quarterly* 105 (1990) 2. Gimbel, "Deutsche Wissenschaftler in britischem Gewahrsam. Ein Erfahrungsbericht aus dem Jahre 1046 über das Lager Wimbledon", in *Vierteljahreshefte für Zeitgeschichte* 38 (1990) 3. Gimbel, "German scientists, United States Denazification policy, and the "Paperclip Conspiracy"", in *The International History Review XII* (1990) 3. Gimbel, *Science, Technology, and Reparations, Exploitation and plunder in postwar Germany* (Stanford: Stanford Press 1990). Hunt, *Secret Agenda, The United States government, Nazi scientists, and project Paperclip, 1945 to 1990* (New York: St. Martin's Press, 1991).

25 Burghard Ciesla, "Das 'Project Paperclip' - deutsche Naturwissenchaftler und Techniker in den USA (1946 bis 1952)", in Jürgen Kocka (ed.), *Historische DDR-Forschung, Aufsätze und Studien*, Band 1 (Berlin: Akademie-Verlag, 1994).

26 Kármán, "Where we stand", I.

27 Michael H. Gorn, *Harnessing the Genie, Science and Technology Forecasting for the Air Force 1944–1986*. Air Staff historical study (Washington, DC: Office of Air Force History, USAF, 1988) 12.

28 Kármán, "Where we stand", II.

29 Gorn, "Harnessing the Genie", 12.

30 Paul A. Hanle, *Bringing Aerodynamics to America* (Boston: Massachusetts Institute of Technology, 1982).

31 The photographs taken on this occasion give the impression of a good relationship between German scientists and AAF-SAG members. *Dryden-Papers, Special Collections*, The Milton S. Eisenhower Library, Johns Hopkins University, Baltimore, Ms. 141, Box 2.4.

32 NASA, Theodore von Kármán 87.1, *U.S.A.F. AECD*, Box 87, Sec VI.

33 BAFB, MF A2055, Frame 0375, 0452, 0575.
34 Ciesla, "Project Paperclip".
35 Names and details of team sizes are only available for the years until 1952. Ciesla, "Project Paperclip".
36 Ciesla, "Project Paperclip".
37 Ciesla, "Project Paperclip".
38 191 specialists were under contract to the Air Force in September 1948. A profession, area of work, or actual job is known for 189 of them. An exact description of the USAF research at Wright's Field cannot now be given, many specialists being given work in a variety of areas. This vagueness allows only an approximate estimate. See: NARA, RG 330, *Specialists now in the United States under the Paperclip Program*, 15. September 1948, Box 15.
39 NARA, R 218, USAF *Report on Project "Paperclip"*, 1947, Box 111.
40 NARA, R 218, Box 111.
41 Calculated according to: NARA, RG 330, Box 15.
42 NARA, RG 218, Box 111.
43 BAFB, MF A2056, frame 0191. *Research and Development Program, Fiscal Year 1947*, AAF Air Technical Service Command (Wright Field, Dayton, Ohio: 1946), 291.
44 NARA, RG 330, *Memo to the director of JIOA, 4 May 1948*, Box 15.
45 NARA, RG 218, Box 111.
46 BAFB, MF A2056, frame 1604.
47 BAFB, MF A2056, *Scientists Program, March 1947* frame 0117. Gimbel, Science, Technology and Reparations, 51.
48 BAFB, MF A2056, frame 0122.
49 BAFB, MF A2056, frame 0122.
50 BAFB, MF A2056, frame 0846.
51 BAFB, MF A2056, frame 0454.
52 BAFB, MF A2056, frame 0456.
53 Dryden-Papers, MS 147, Box 2.40.
54 NASM, Theodore von Kármán 87.1, *U.S.A.F. AECD*, Box 87, Sec. VI. Notes of conversations and interview with Dr. Hiebert, Chief Historian of AEDC, Tennesseel Dr. Heinrich Ramm, former Paperclip scientists; Dr. Eastman, former scientific director of AEDC. These interviews took place on 30 April 1992 in the AEDC, Tennessee.
55 Horst Mönnich, *BMW, Eine deutsche Geschichte* (München/Zürich: Piper Verlag, 1991) 359.
56 Hans-Joachim Braun/Walter Kaiser, *Energiewirtschaft-Automatisierung-Information seit 1914, Propyläen Technikgeschichte, Band 5* (Berlin: Rowohlt, 1993) 9–59.
57 Dieter Hoffmann (ed.), Operation Epsilon, Die Farm-Hall-Protokolle oder Die Angst der Alliierten vor der deutschen Atombombe (Berlin: Rowohlt, 1993) 9–59.

58 Rolf Kreibich, *Die Wissenschaftsgesellschaft, Von Galilei zur High-Tech-Revolution* (Frankfurt/M: Suhrkamp, 1986), 336.

59 Anselm Franz "Der Weg zum ersten Großserien-Strahltriebwerk Junkers Jumo 004 und spätere Entwicklungen in den USA", in *50 Jahre Turbostrahlflug* (Bonn: DGLR, 1989) 121.

60 Ciesla "Der Spezialistentransfer in die UdSSR und seine Auswirkungen in SBZ und DDR", in *Aus Politik und Zeitgeschichte*, supplement to "*Das Parlament*" (No. B 49–50/93, 3 December 1993).

61 Werner Link "Die amerikanische Deutschlandpolitik 1945–1949" in *Studien zur Deutschlandfrage, Band 7* (Berlin: Duncker und Humblot, 1984), 7–8.

62 Christoph Buchheim, Die Wiedereingliederung Westdeutschlands in die Weltwirtschaft 1945–1958 (München: R. Oldenbourg Verlag 1990) 77.

CHAPTER 8

Immaterial Reparations and the Reintegration of West Germany into the World Market

Werner Abelshauser

"Dismantling and decartelization were economic sanctions which ultimately contributed to the German steel industry's long years of superiority to the steel industries of other nations — a lesson of history"[1]

(Hans-Günther Sohl)

I

It was a long time before German industrialists spoke publicly in a positive vein of the effects of reparations following World War II, and particularly of the dismantling of West German industry. Until well into the 1960s, complaints of the persistent restrictive effects of dismantling on start-up chances for West German industry were part of the standard repertoire of special interest lobbying by the relevant associations. For a long time, they were also supported by historians, who saw in the dismantling "severe damage to an economy shattered by the effects of war."[2] Not until the 1970s did a more discriminating analysis of the dismantling prevail. In an almost complete reversal of the previous assessment, most historians now saw the dismantling of industrial capacity by the occupation forces and the Inter-Allied Reparations Agency as a virtual prerequisite for the "economic miracle" because the material reparations made by the West German economy in the 1950s had opened the way for rapid renovation and streamlining of the industrial capital stock.[3]

At the same time, there were other reasons for questioning the significantly negative effect of dismantling on West German reconstruction and the return of the Federal Republic to the world market. Economic history research furnished evidence that the scope and structure of the dismantling could not have had a significant effect on economic development in the early postwar period (Table I).[4] Although both positions call into question the significant negative effects of reparations feared by the people at the time, in the last analysis they are mutually exclusive. They relate almost exclusively, of course, to material reparations, i.e., to the dismantling of the industrial capital stock, to the confiscation of German foreign assets and to restitution of industrial facilities looted by the National Socialists. More recent research,[5] however, has pointed out the importance of so-called intellectual reparations, i.e., the confiscation of patents, production documentation, industrial secrets, etc., and has accorded them a greater significance than material reparations. In the light of this research, and in view of the changing focus of interest in "dismantling research," the question can be posed again whether the reparations policy of the Allies led, at least from this perspective, to significant impairment of West German reconstruction and West German export trade.

For various reasons, it is extraordinarily difficult to answer this question. Practically insurmountable obstacles arise even in deciding how to approach the question. A quantitative approach to immaterial reparations quite obviously represents a contradiction in terms. At any rate, no convincing method has been found to date to express in *economically relevant* numbers the immaterial damage which indisputably occurred as a result of the confiscation of industrial

Table I Structure of dismantling in West Germany

Branch of Industry[a]	Percentage of Dismantling Value[b] up to the End of 1948 (%)	Growth in Real Gross Production in the Third Reich 1936–1944 (%)
Machine building	37.4	
Aircraft construction	16.1	+143.1 (Machine building, Aircraft construction, Shipbuilding)
Shipbuilding	1.8	
Chemistry and fuels	35.5	+ 72.9
Iron and steel	3.5	+116.9
Nonferrous metals	2.9	+ 38.7 c
Energy supply	1.7	+ 77.0 d
Precision mechanics	1.1	+ 91.4

[a] The plants declared by the Inter-Allied Reparations Agency (IARA) to be "pure" armaments operations were included in the different industry branches. According to IARA data, they comprise about 70% of the total value of dismantled property;

[b] Remaining value in 1938 prices according to the valuation guidelines of IARA;

[c] Iron and steel producing, non-ferrous metals and foundry industries;

[d] 1936–1943.

Calculated based on: Agence Interalliée des Reparations, *Rapport du Secretaire General pour l'année 1949* (Brussels, 1950), p. 23; B. Gleitze, Ostdeutsche Wirtschaft (Berlin, 1956), p. 169; Verwaltung für Wirtschaft, "Die Wirtschaftslage der vereinten Zone," (Sept.–Oct. 1947) BA Z 8/207.

secrets, patents and other production documentation. The estimates, as reported for example by Jörg Fisch,[6] vary between 100 million marks and $10 billion. The "calculations" on which these different numbers are based are educated guesses at best. Most are merely ranges pulled out of the air to support the positions of certain interest groups. Since this is a fundamental problem, better figures can hardly be expected from further research.

We must therefore rely on the qualitative approach to determine the economic significance of immaterial reparations. Paradoxically, however, for theoretical and logical reasons, this is possible only if *immaterial* reparations are seen in close relationship with the *material* dismantling, because they are inseparable from the possible subjects or embodiments of technical and/or organizational knowledge, and they gain economic relevance only in this connection. Specifically, restrictive effects on reconstruction and export are measurable only if immaterial reparations led to the construction of competing capacities in the countries which received the reparations, and if this capacity replaced German exports to those countries, or displaced German exports by direct competition in third-party markets. Restrictive consequences can also be demonstrated, however, if immaterial reparations together with dismantling completely pulled the foundation out from under the export trade in certain branches of industry, i.e., if entire branches which were significant for German industry before the war, such as the armaments industry, completely ceased to exist.

Precisely the example of the armaments industry, which was in reality not insignificant in the 1950s, also shows that the transfer of qualified workers is another important factor which must be included when analyzing the effects of intellectual reparations. This brushes against the boundary of what is covered by the term "reparations," however. While the transfer and use of workers in the mines of the Soviet Union and France can be easily classified as reparations, the "brain drain" of German scientists, technicians, engineers and other experts to these countries and also to Great Britain and the United States after 1945 was not always involuntary. While around 1,000 German scientists and experts were brought to the US in the context of Operations "Overcast" (July 1945) and "Paperclip" (1946/47), the real emigration of German scientists did not begin until quite a bit later. Between 1949 and 1966, about 1,800 scientists and approximately 4,200 technicians emigrated to the United States.[7] We can therefore speak of "intellectual reparations" only if the "recruitment" of German workers immediately after the war was pursued systematically and with a certain vigor. This applies not only to the aforementioned US immigrants of the early postwar period, including such prominent persons as Werner von Braun and his rocket experts, but also to the 1,000 or so German experts who were brought to Great Britain by BIOS (British Intelligence Objectives Sub-Committee) in the context of British postwar planning in the area of industrial policy.[8] The concept of "intellectual reparations" can obviously not be applied to the majority of cases, where the transfer occurred on a more or less "private" basis. Therefore, in terms of the economic effects which intellectual reparations and later enticement of workers out of Germany exerted on German reconstruction and return to German exports to the world market, the transfer must be differentiated by country and by branch of industry. The former affects the competition which the German economy faced in the postwar period; the latter permits conclusions regarding the relevance of intellectual reparations for reconstruction of the German economy. For example, the Soviet Union represented only minimal competition for German exports, and the drain of armaments experts should have had no negative effect on German economic development in the 1950s.

It is extraordinarily difficult to prove the restrictive effect of this "brain drain" on German exports in all other cases, however. Technological knowledge is not just stored in patents, not just materialized in machines and not just embodies in the workers. Rather, it is incorporated institutionally and organizationally into a national economy, i.e., it can be reproduced and made economically available again as long as there is a functional educational and training system and the basic economic structure remains intact.[9]

II

These are the most important ways in which research would have to approach this subject. The considerations presented in the first chapter show that the value of patents, industrial secrets and production documentation, even if it could be determined precisely, is of only very limited usefulness as an indicator of the extent of restrictive effects of immaterial reparations on German export trade. This approach would be conclusive only to the extent that it related to the export of know-how in a narrower sense, i.e., cross-border trade in patents, licenses and the like.

Apparently of much greater significance, however, is the forced technology transfer from Germany to the countries which were reparations recipients, to the extent that it contributed to development there of competitive export capacities. According to the research to date, the most important example for this correlation between immaterial reparations and export trade would appear to be the case of Great Britain. Carl Glatt has demonstrated in his dissertation[10] that, even before the end of the war, ambitious plans were being laid in Great Britain for a systematic expansion of British export capacity at the expense of the German market share. The prerequisite for taking over this market share was to be created by transplanting export-relevant technologies from German to Great Britain. This applied above all to those branches of industry where development would enable the British to replace imports from Germany, such as the textile industry (especially hosiery), the clock and watch industry, the glass industry and the manufacturing of photographic materials. Also included were heavy-export industries such as optical instruments, dyes and pharmaceuticals, petrochemical products, plastics, phonograph records and machine tools, with which Great Britain hoped to strengthen its own position in the world market. The transfer of German technology to Great Britain was also intended to provide positive impulses for the use and ongoing development of the supposed "cutting edge technologies," particularly in the armaments and aviation industries. The heart of this concept was the transfer of key technologies in the form of written production documentation, machine prototypes and (parts of) industrial plants, as well as the transfer of qualified workers and scientists, who were indispensable for the practical implementation of these new technologies. Between the autumn of 1944 and June 1947, therefore, more than 12,000 British and American investigators combed through German companies and research institutes and confiscated approximately 10,000 documents.[11]

"Operation Plunder," as this campaign of the British "T" Force Army Group was suggestively called, was apparently not without success, and the calculations of the British industrial planners were met at least in part, as signalled by data on the declining German import share of important categories of goods during the 1950s. For example, before the war, about 30% of the flat glass used in Britain had to be imported, with more than 10% of the imports coming from Germany. After the war, Great Britain rapidly developed worldwide into a leading export nation in this sector. The development in the clock and watch industry was similarly dramatic. In 1937, the British had manufactured almost no small time pieces. By 1951, on the other hand, imports had essentially been replaced with domestic production, and Britain's own exports had grown to 1,350,000 pounds Sterling. Corresponding to this development, West Germany's share in the world market for clocks and watches shrank from 11.7% (1937) to 9.2% (1951),[12] not least of all due to the loss of the British market. Other examples in the optical and chemical industries, as well as numerous smaller industries, could also be named here. One of the positive results of this postwar industrial policy was, without a doubt, the fact that Great Britain did not fall behind the other developed industrial nations in all areas of industrial production as it did, for example, in coal mining and in the textile and automobile industries. The British economic successes included not only the glass industry, pharmaceutical industry, aircraft construction, petrochemicals and plastics, and certain specialized areas of machine building and the phonograph record industry, but also a surprising number of sectors which were at the center of British efforts to transfer technology out of their German occupation zone. If it can thereby be demonstrated that a positive effect for the British postwar economy — however limited — resulted from the immaterial reparations from Germany, it does not mean, of course, that this was associated with a simultaneous, significant disadvantage to the West German economic recovery and export capacity.

This is demonstrated by the fact that Great Britain by no means lost its important overall role as an export market for German capital and consumer goods in the 1950s, notwithstanding the individual cases mentioned above. Using the example of the "Little Marshall Plan," it can even be shown that West German industry regained its ability to compete with British industry as early as the late 1940s. Under the two intra-European currency and clearing agreements signed under the aegis of the Marshall Plan (1948/50), West Germany soon developed a trade surplus and became an attractive supplier to the British economy. It was the intent of this "Little Marshall Plan" to stimulate intra-European trade and, to this end, to force the countries with a potential intra-European trade surplus to grant drawing rights in the form of foreign currency to their trading partners who had trade deficits. For the countries with a trade surplus, therefore, aid dollars were converted into "conditional aid," i.e., they were granted only if they were passed on as drawing rights in their own currency to the countries with trade deficits. While German importers did not take advantage of the drawing rights granted to them in British currency because British goods and services were not competitive with the German offerings, West Germany itself had to grant its neighbors more drawing rights than had originally been planned.[13] Although this reduced the West Germany share of material Marshall Plan aid, West Germany was able in this way to reestablish links to old markets at this early date and simultaneously demonstrate its export capacity.

It is true that Great Britain continued to develop its leading position in Europe in the areas of aviation technology and the armaments industry in the 1950s and thereby profited to a certain degree from the intellectual reparations from Germany. It is more than a little doubtful, however, that this damaged German economic interests. Rather, there are indications that it was precisely this specialization that impaired the chances and possibilities for British export in other areas because it tied up a large portion of the available capacity.[14] Even the goal of import replacement, toward which Great Britain was able to move in some areas using German know-how, was not necessarily in the considered interest of the British economy — even if it did lead to a superficial easing of pressure on the balance of payments. For example, after 1945, British industry finally succeeded, with the help of German techniques and technicians, in manufacturing jewelled parts for watch production (sapphire boules), but at a price that was four times that on the world market. This is because the solution to the problems of horological jewel-making had less to do with the availability of suitable machines and descriptions of the manufacturing process than with the long years of experience of the diamond cutters. However, the responsible authorities and unions, as well as the British public, erected barriers to the transfer of a large number of German specialists.[15] All of these factors contributed to the reopening to West German exports of the markets which had previously been lost to Great Britain. At any rate, the argument that German exports enjoyed uninterrupted competitiveness is supported by the fact that West Germany was able to raise its share of world exports in the 1950s from 3.6% (1950) to 8.9% (1960), while Great Britain dropped from 11.0% to 7.8%.[16]

The extent to which the effects of intellectual reparations are responsible for German exporters' difficulties in returning to the North American market must be investigated in detail. The complete loss of patents and trademarks after the London Agreement of June 27, 1946, which suspended protection for all former German patents and made them available to any interested party free of charge, was certainly not without its negative effects. It was not without reason that the release of German trademarks and copyrights, along with the issue of confiscated German assets, remained a continual subject of negotiations between the German and American governments well into the 1960s. A definitive end to the expropriation of German assets in the US under the "trading with the enemy act" was not reached until April 1953.[17] It was not until 1956, however, with the signing of the Patent Interchange Agreement, that legal security was restored to the technology transfer with the United States. The other problems with German exports to the United States, however, lay more in the tariffs and administrative barriers which a still vital US protectionist tradition erected against German and other exporters.[18] Nevertheless, German exports managed to hold onto a long term share of about 6–7% of the North American market.

It is certainly too soon to draw definitive conclusions regarding the effects on German reintegration into the world market exerted by German technology transfer into the countries which were the recipients of immaterial reparations. As regards the possible effect of replacement of German imports or competitiveness in third-party markets, however, the answer would tend to fall in the positive column. Under the expansive conditions of the 1950s, the reparations-related "sell-out" of German technology and German experts to Germany's main trading partners probably led less to a restriction of German market chances than to their improvement. In many cases, namely, the forced technology transfer led to establishment or deepening of relationships between German and foreign companies.[19] Characteristics of the integrative character of West German exports in the 1950s was their direct entwinement with the export trade of Germany's European neighbors. This was also the foundation of the extraordinary dynamism of German export trade. These neighbors received German finished products and also supplied semi-finished goods which were used in German export products, e.g., as machine units. German export trade thus became an important market for European producers and the creation of a common market was anticipated even before the founding of the European Economic Community. In many cases, presumably, this development arose not from trade policy strategies but from the schemes forming the history of reparations.

III

The clearest rebuttal is to the question of whether West German integration into the world market was hampered by the prohibition against use of technologies in connection with the dismantling of entire branches of industry. Table I shows that the structure of West German industry was not significantly changed by the dismantling, aside from "pure" armaments industries, which were subject to production bans until the mid-1950s and, in some cases, beyond. These restrictions do not appear to have been perceived as burdensome by West German industrialists, however. At any rate, there was markedly little eagerness within West German industry to rebuild this area of production in the early 1950s.[20] In many cases of armaments-induced technologies, e.g., in the synthesis of rubber (Buna) or gasoline (Fischer-Tropsch method), the production ban was irrelevant simply because the price situation on the world market did not permit economical use of these new technologies. Of course, this argument cannot be generalized to all cases. Some bans and restrictions definitely led to concrete impediments, even in relevant markets. Examples of such bans are those imposed by Law No. 22 (March 15, 1950) and Law No. 24 (March 30, 1950) of the Allied High Commission. The former dealt with nuclear reactors and particle accelerators, and the latter with products of steel and rolling mills, chemistry, ship-building, electrical technology and machine tools. For example, permission was required for construction of x-ray facilities over 150,000 volts, although 250,000-volt instruments were the technical standard in 1950. Another example is the ban on production of electron tubes over 250 megahertz, although these were required for the manufacturing of television sets.[21] However, even before they were finally lifted in the Paris Treaties of May 1955, many of these restrictions, controls and bans lost their significance in light of the Korean War.

Other data (Tables II and III) illustrate that the immaterial components of German industrial capital stock, i.e., its technological standard and resulting efficiency ("quality") were considerable even after the end of the dismantling compared to other time periods. It is nearly impossible in this context to construct a restrictive correlation between intellectual reparations and German export trade, particularly since the growth rate in exports in the 1950s was significantly greater than that of the economy overall.

For the same reason, one would be hard pressed to accept the naive assessment that it was the dismantling that (first) cleared the way for streamlining and modernization in German industry. The age breakdown, efficiency and technological standard of the branches of industry relevant to export trade provide no evidence of a generally backward state of development which had to be overcome before the export offensive could succeed. Intellectual reparations notwithstanding, 80% of the brand name exports from Germany in 1952–53 used trademarks which had been valid both domestically and abroad before the end of the war.[22] However, even the West German export industry could not have survived past the medium term without modernizing and streamlining its production apparatus and product offerings. In this respect, the dismantling, even the confiscation of intellectual property, had a decisively positive effect on the conditions of development for German industry, though in a different sense than historians usually assume. Namely, the argument of damages suffered from dismantling was masterfully handled from the beginning by industrial interest groups in order to promote their own ideas for economic, fiscal and — especially — tax policy before the state and federal governments. Their goal was not so much to obtain actual loans to undo the dismantling, such as those which the state of North Rhine-Westphalia granted as early as 1948 to industries damaged by dismantling. Instead, the primary purpose was to gain advantages in the handling of amortization for tax purposes, property assessments in the opening Deutschmark balance and other tax

Table II Age distribution of the gross investment in fixed assets of West Germany industry in 1935, 1945 and 1948 (in %; total of all age categories = 100)

<table>
<tr><th>Age Categories (years)</th><th>1/1/35</th><th>1/1/45</th><th>1/1/48</th></tr>
<tr><td colspan="4">Mining, primary commodity and producer goods industries</td></tr>
<tr><td>0–5</td><td>9</td><td>40</td><td>18</td></tr>
<tr><td>5–10</td><td>21</td><td>24</td><td>40</td></tr>
<tr><td>10–15</td><td rowspan="2">70</td><td>5</td><td>12</td></tr>
<tr><td>16 +</td><td>31</td><td>30</td></tr>
<tr><td colspan="4">Capital goods industry</td></tr>
<tr><td>0–5</td><td>7</td><td>37</td><td>19</td></tr>
<tr><td>5–10</td><td>15</td><td>20</td><td>32</td></tr>
<tr><td>10–15</td><td rowspan="2">78</td><td>5</td><td>12</td></tr>
<tr><td>16+</td><td>38</td><td>37</td></tr>
<tr><td colspan="4">Consumer goods industry</td></tr>
<tr><td>0–5</td><td>10</td><td>18</td><td>10</td></tr>
<tr><td>5–10</td><td>22</td><td>16</td><td>21</td></tr>
<tr><td>10–15</td><td rowspan="2">68</td><td>8</td><td>12</td></tr>
<tr><td>16+</td><td>58</td><td>57</td></tr>
<tr><td colspan="4">Food industry</td></tr>
<tr><td>0–5</td><td>12</td><td>21</td><td>13</td></tr>
<tr><td>5–10</td><td>23</td><td>18</td><td>22</td></tr>
<tr><td>10–15</td><td rowspan="2">65</td><td>10</td><td>14</td></tr>
<tr><td>16+</td><td>51</td><td>51</td></tr>
<tr><td colspan="4">Small-scale industrial enterprises</td></tr>
<tr><td>0–5</td><td>9</td><td>15</td><td>9</td></tr>
<tr><td>5–10</td><td>15</td><td>14</td><td>17</td></tr>
<tr><td>10–15</td><td rowspan="2">76</td><td>9</td><td>12</td></tr>
<tr><td>16+</td><td>62</td><td>62</td></tr>
<tr><td colspan="4">Industry overall[a]</td></tr>
<tr><td>0–5</td><td>9</td><td>34</td><td>16</td></tr>
<tr><td>5–10</td><td>15</td><td>21</td><td>34</td></tr>
<tr><td>10–15</td><td rowspan="2">76</td><td>6</td><td>12</td></tr>
<tr><td>16+</td><td>39</td><td>38</td></tr>
</table>

[a] Not including energy supply and construction industry.

Source: DIW [German Institute for Economic Research], *Sonderhefte*, N.F., No. 42 (Series A), pp. 52f.

advantages such as the granting of payment extensions, grace periods, etc. It is impossible to exaggerate the weight of the dismantling argument in implementing this policy of tax and fiscal privileges when one considers that independent financing was by far the most important funding mechanism in the 1950s. The

Table III Quality[a] of the fixed assets of West German Industry, 1936–1948

Main groups/year[b]	1936	1945	1946	1947	1948	1949
Mining, primary commodities and producer goods	49.0	63.7	61.6	59.6	57.9	56.3
Capital goods	47.6	62.8	61.0	59.0	57.6	57.4
Consumer goods	54.1	54.9	53.8	53.1	52.6	53.7
Food	52.5	55.1	53.7	52.7	52.1	53.3
Small-scale industry	50.7	50.9	50.0	49.3	48.9	49.1
Industry overall	49.9	61.3	59.4	57.7	56.3	55.7

[a] Relation of net fixed assets to gross fixed assets (= Gütegrad):
[b] On January 1 of each year.
Source: DIW [German Institute for Economic Research], *Sonderhefte*, N.F., No. 42, p. 79.

goal of the relevant interest groups, e.g., the Notgemeinschaft der reparationsgeschädigten Industrie (NRI) [Emergency Association of Reparations-Damaged Industry], therefore, was not so much to prevent material or immaterial reparations, aside from individual disputed properties, as to establish the instruments for defense against dismantling in order to achieve reconstruction goals which were important for reasons of industrial and capital market policy, and to eliminate economic constraints to ensure that economic growth happened according to the free will of industry.[23] It is in this sense that we should understand the triumphant position of Hans-Günter Sohl (Thyssen Steel) which was quoted at the beginning of this paper.

It is certainly too early to draw a definitive conclusion on the issue raised here. However, all indications are that possible restrictions fall in an early period of economic reconstruction in which other restrictions and barriers far outweighed any conceivable negative effect of the forced technology transfer. This applies particularly to the long years of restrictive effects of the occupation's control over German export trade, the so-called "Dollar Clause", production bans, etc. In actuality, the effect of Operation Paperclip was only one of numerous factors impeding rapid recovery of the West German economy after its collapse. In cases where real disadvantages arose for German export companies as a result of immaterial reparations, these disadvantages were generally more than compensated very soon by the blessings of world market expansion. It seems a worthy research goal to ask whether the forced technology transfers — and also the brain drain of German scientists and technicians after World War II — did not significantly help to increase the entwinement of the German export industry with its most important foreign markets and thus lay an important foundation for the success of German exports in the 1950s.

Translated by Sally Robertson

NOTES

1 Hans-Günter Sohl, "Nur die Sanktionen machten den Aufstieg der Stahlindustrie möglich" *Welt am Sonntag* (June 4, 1989) *23*: 44.

2 Wilhelm Treue, Die Demontagepolitik der Westmächte nach dem Zweiten Weltkrieg unter besonderer Berücksichtigung ihrer Wirkung auf die Wirtschaft in Niedersachsen (Göttingen/Frankfurt/Zürich: Musterschmidt, 1967), p. 89 is a typical example.

3 For example, Wolfgang Benz, "Wirtschaftspolitik zwischen Demontage und Währungsreform", in: *Westdeutschlands Weg zur Bundesrepublik 1945–1949*, ed. Institut für Zeitgeschichte [Institute for Contemporary History] (Munich: R. Oldenbourg, 1976), pp. 67–89.

4 Werner Abelshauser, *Wirtschaft in Westdeutschland* (Stuttgart: Deutsche Verlagsanstalt 1975) pp. 114–129.

5 For example, John Gimbel, *Science, Technology, and Reparations. Exploitation and Plunder in Postwar Germany* (Stanford, CA: Stanford University Press 1990).

6 Jörg Fisch, *Reparationen nach dem Zweiten Weltkrieg* (Munich: C.H. Beck 1992), pp. 213f.

7 Thomas Stamm, *Zwischen Staat und Selbstverwaltung. Die deutsche Forschung im Wiederaufbau 1945–1965* (Cologne: Verlag Wissenschaft und Politik 1981) p. 45; Jean-Jacques Servan-Schreiber, Die amerikanische Herausforderung (Hamburg, 1968), p. 16.

8 Carl Glatt, *Reparations and the Transfer of Scientific and Industrial Technology from Germany: A Case Study of the Roots of British Industrial Policy and of Aspects of British Occupation Policy in Germany between Post-World War II Reconstruction and the Korean War, 1943–1951* (Dissertation forthcoming at the EUI Florence, 1994).

9 Werner Abelshauser and Dietrich Petzina, "Krise und Rekonstruktion. Zur Interpretation der gesamtwirtschaftlichen Entwicklung Deutschlands im 20. Jahrhundert", in: *Historische Konjunkturforschung*, ed. by Wilhelm Heinz Schröder and Reinhard Spree, (Stuttgart: Klett-Cotta 1980), pp. 95–114.

10 *op. cit.*

11 See, especially, Gimbel, *op. cit.* and Glatt, *op. cit.*

12 Henri Rieben, *et al.*, *L'Horologerie et l'Europe* (Lausanne, 1959), p.42.

13 Werner Abelshauser, "The re-entry of West Germany into the international economy and early European integration," Western Europe and Germany. The Beginnings of European Integration 1945–1960, ed. by C. Wurm, (Oxford, Washington, 1995), pp. 27–53.

14 This has been suggested by, among others, Correlli Barnett, *The Audit of War. The Illusion and Reality of Britain as a Great Nation* (London: Macmillan 1986), Malcolm Chalmers, *Paying for Defence* (London: Pluto Press 1985) and Andrew Shonfield, *British Economic Policy since the War* (Harmondsworth, 1958).

15 This example is cited in Glatt, *op. cit.*, Chapter 1.
16 United Nations Statistical Yearbook, *passim*.
17 Hans-Dieter Kreikamp, *Deutsches Vermögen in den Vereinigten Staaten. Die Auseinandersetzung um seine Rückführung als Aspekt der deutsch-amerikanischen Beziehungen 1952 bis 1962* (Stuttgart: Deutsche Verlagsanstalt 1979).
18 Reinhard Neebe, *Überseemärkte und Exportstrategien in der west-deutschen Wirtschaft 1945 bis 1966* (Stuttgart: Franz Steiner Verlag 1991).
19 Indications of this are found in both Raymond G. Stokes, *Divide and Prosper. The Heirs of IG Farben under Allied Authority 1945–1951* (Berkeley: University of California Press 1988), pp. 189ff. and John Gimbel, *op. cit.*, pp. 183f.
20 This was the impression obtained by the US consul in Düsseldorf after conversations with Otto Wolff von Amerongen and West German steel managers. AMCONGEN, Düsseldorf to Department of State on December 22, 1955, NA762A.5/12–2255 and on December 13, 1955, NA762A.5/12–1355.
21 Reinhard Neebe, "Technologietransfer und Außenhandel in den Anfangs-jahren der Bundesrepublik Deutschland", in: *VSWG* (1989) *76/1*: 51f.
22 Ludwig Erhard, Deutschlands Rückkehr zum Weltmarkt (Düsseldorf, 1953), p. 255.
23 An important finding of the as yet unpublished dissertation of Martina Köchling, Demontagepolitik und Wiederaufbau in Nordrhein-Westfalen (Bochum: Ruhr University 1992).

CHAPTER 9

The Return of German "Specialists" from the Soviet Union to the German Democratic Republic: Integration and Impact

André Steiner

Public knowledge of the work of German "specialists" in the USSR after World War II was, for a long time, largely limited to the memoirs of the more prominent scientists and engineers who had participated in Soviet armaments programs.[1] The opening of the German Democratic Republic's (GDR) archives now provides access to new sources for assessing the extent of this intellectual transfer. The present paper, however, focuses neither on how these technical experts were "recruited" nor on their subsequent activities in the Soviet Union. Initial research findings in these areas were recently published by Ulrich Albrecht *et al.*[2] The aim here is to make use of the files in the archives of the former GDR in order to reconstruct the return of these specialists to East Germany and their impact. In a parallel project, Burghard Ciesla has conducted a study of the card file of the GDR's Ministry of the Interior (*Ministerium des Inneren; MdI*). It was possible to incorporate our first joint evaluation of these documents into the original version of this paper, which was presented in Washington.[3] As in that presentation, "specialists" will be used here to designate not only top scientists, but also those engineers, technical personnel, master craftsmen, and skilled laborers required in the practical application of scientific findings.

The specialists taken to the USSR after the end of World War II basically began to return in 1950. From the beginning, central state authorities attached great importance to integrating them properly into the GDR. The "first special operation" (*erste Sonderaktion*) brought the first large group of technical experts, transported in five subgroups and arriving between September 19 and October 10, 1950. This operation included about 400 specialists and their families, or a total of about 1,000 people. They were brought to the resettlement camp in

Wolfen, where they were to stay under they were found employment. The majority of these specialists had previously lived in Dessau and were thus primarily involved in aircraft construction.[4] A second wave of returnees in December 1950 brought back a total of somewhat more than 200 specialists.[5] Thus, between September 19 and December 31, 1950, 610 specialists and 1,078 members of their families were received in Frankfurt on the Oder.[6] By the middle of February 1951, the number of specialists returning to the GDR had reached 668. Most had apparently been employed in aircraft construction. Their occupations and qualifications ranged from that of skilled laborer (especially from the metal industry) or craftsman, to that of technically trained white-collar employee, all the way to that of physicist, aeronautical specialist, mechanical engineer, or design engineer.[7] This occupational diversity created a series of difficulties in integrating the returnees, even though they were urgently needed in industry.[8] Thus, the Minister of Labor and Health Services of the state (*Land*) of Sachsen-Anhalt, Becker, complained that "It is very difficult to place some of these individuals, since they are very specialized. For example, one of them is only a test pilot (*Einflieger*) for jet fighters. He was not trained in any special occupation. Previously he had flown test flights in Germany and recently in the Soviet Union."[9]

It took time to find positions for the specialists, since the "remigrants" negotiated with several authorities at once. Their high salary demands led to further delays in concluding contract negotiations. It proved particularly difficult to find suitable positions to offer to former executive personnel. The enterprises in East Germany only had positions with monthly incomes of DM 500–600, but the specialists demanded DM 800 and upwards.[10] Most negotiations involved the general machine-building industry, where in 1950 the average monthly income for employees was DM 271.[11] According to GRD standards, the salary demands of the returnees were strikingly high. They, however, considered them entirely justified, since they had already been paid relatively high salaries in the Soviet Union, both in absolute terms and relative to the incomes earned there. Most were aware of their value and were convinced they could find a suitable new position. On the other hand, the enterprises had little interest in hiring employees who would earn above-average incomes: they wanted to maintain their internal wage structure and they were only allotted limited funds for covering wage costs. For this reason, the high-priority political objective of getting the specialists to commit themselves to the GDR had to lead to conflicts of interest between the enterprises and the central political authorities. And these conflicts could only be resolved by administrative intervention on the part of the central authorities in their own regulatory codes. In individual cases involving some of the more outstanding scientists or engineers who had already returned, special individual contracts (*Einzelverträge*) were made use of. This was a normal form of special employment contract in the GDR and entailed various social privileges. Necessary funding was directly provided by the central authorities.

According to an investigative report on the specialists who returned in 1950 to the key areas of Berlin, Dessau, Bernburg, and Merseburg-Leuna, they in part managed to evade centrally organized supervision. Nevertheless, they held positive views of the political development in both the GDR and the Soviet Union. This points to the fact that of the technical experts who returned to the GDR, only those remained who were not fundamentally opposed to the system. The rest preferred to move to the western part of Germany. Sixty per cent of those placed by the State Skilled Employment Offices (*Facharbeitsämter*) in Berlin were able to reoccupy their earlier positions. But all the rest were also employed in occupational groups in keeping with their expertise. In the fields investigated, 125 specialists had positions in "Soviet corporations" (*Sowjetische Aktiengesellschaften; SAG*), 220 were employed by state enterprises, and 3 worked in private firms.[12] The insignificant representation in private industry, which at that time still employed 32.3% of the entire industrial work force[13], was the result of the great interest of central state authorities in employing highly qualified laborers in state enterprises or in the privileged SAG realm.

This first wave of repatriated German specialists from the Soviet Union apparently produced a certain degree of euphoria among those still remaining in the USSR. This was received rather skeptically by the responsible GDR authorities, the Central Office of Research and Technology (*Zentralamt für Forschung und Technik; ZAFT*) at the State Planning Commission (*Staatliche Plankommission; SPK*).[14] And their doubts turned out to be completely justified. Whereas in 1950 alone, 24.3% of all specialists returned, only 9.7% more came in 1951.[15] Many of the 70 chemical experts who returned in the spring of 1951 were simply able to return to their former positions at the *Farbenfabriken Wolfen*, the *Elektrochemisches Kombinat Bitterfeld* and, the *Leuna-Werk*. Both the SAG and state enterprises provided material support and company apartments. Specialists from industries which were forbidden in Germany, such as aircraft construction, however, had to be placed in other, "related" industries, such as shipbuilding. In these cases, it was usually not possible to immediately assign the returnees a new position. Some were also given employment in central administrative positions or at the Association for German-Soviet Friendship. In the general estimation of the MdI, the qualifications and experience of the specialists were taken into account as far as possible in the assignment of a position.[16]

The procedures for reintegrating the specialists were, however, the cause of some controversy among the central authorities. ZAFT stated that inadequate organization had produced considerable difficulties in the first special operation "in assigning the in part extremely valuable specialist ... with suitable positions. The supervision of the experts suffered furthermore from the fact that no single authority was given overall control and thus supervision of proper accommodation [of the specialists with positions] could not be implemented." Accordingly, the director of ZAFT recommended that the responsibility for finding employment for the experts be put under the authority of his office.[17] Apparently, however, ZAFT only had the more prominent scientists and engineers in mind and paid less attention to the majority of those already arrived, who were skilled laborers and master craftsmen. For this reason, the MdI was of the opinion that the Ministry of Labor should take on responsibility for career counseling and job placement.[18] Starting in December 1950, this ministry was informed by the MdI 2–3 days prior to the arrival of a "transport." Since, however, the fields of work of those arriving were not known, surveys of need were conducted in the state enterprises. The personnel directors were obliged to employ specialists from the Soviet Union assigned to them in accordance with their qualifications. In this way it was possible, in coordination with the Ministry of Machine-Building Industries (*Ministerium für Maschinenbau; MfM*), to provide the 174 specialists who arrived in December 1951 with 1,000 positions to choose from. These experts — largely skilled laborers from the metal industry — were counseled and chose, with few exceptions, the positions recommended to them, a success which was likely the result of the relatively large pool of positions offered. Those arriving certainly appreciated the efforts made to supply them with positions and apartments. Nonetheless, they also considered this only natural. Starting at the end of 1951, though responsibility remained in the hands of the Ministry of Labor, the industry-specific ministries, and here, especially the MfM, were supposed to be more intensively incorporated into the process. They now had to assess need and supervise subsequent employment. Thus, the specialists were now subject to three-fold supervision: the enterprises registered any employment negotiations with the MfM; the responsible authorities of individual state governments monitored the provision of apartments and positions; and the specialists themselves had to register their new place of residence when they picked up their transferred monies at the German Central Bank (*Deutsche Notenbank*).[19]

Nevertheless, the overlapping areas of authority continued to leave the impression of improvisation, an impression which was in part accurate. This prompted the director of ZAFT to call once again for the concentration of all authority in his office. Since those returning knew little about the development of the GDR and its politics and the most capable of the specialists were convinced of their abilities to immediately find new positions in West Germany, he considered it necessary to create an "atmosphere of trust." In the view of ZAFT, the only method that had proven itself up until then was the preferential provision of apartments. In contrast, the conditions in the camp at Wolfen were severely criticized. Several families had to be accommodated in each room of the unadorned barracks. The sanitary conditions ranged from "completely inadequate" to "primitive." The staff employed in career counseling was "only [qualified] for counseling craft specialists. ... These conditions often resulted in the homecoming specialists traveling around the GDR for several months and still not ending up in the most suitable position. Thus, there are unfortunately many known cases where returning specialists have migrated to the West. In this sense, the problem also possesses considerable political importance." Moreover, there were also formal, bureaucratic barriers against the implementation of individual contracts (*Einzelverträge*), since the ministries had in part exhausted the allotments provided by the Ministerial Council (*Ministerrat*) and had first to apply for additional funds. There was a real reason, however, why ZAFT now pushed for a quick solution to the problems at hand: though most of those who had return until now had been skilled laborers and technical personnel, the outstanding scientists and engineers who had mostly been employed in armaments research were now awaited. For this reason it was stressed that: "For reasons of secrecy, supervision should be carried out from one central authority, since it is in expedient and inadmissible that complete lists of returnees should exist in more than one place." ZAFT once again recommended itself as the professionally qualified authority for the task.[20]

In 1952, 31.2% of the specialists returned. This meant that almost two-thirds of them were now repatriated.[21] Starting in 1953, the recommendations of ZAFT were in part realized. Overall coordination still remained the task of MdI, precisely because now greater numbers of more well-known scientists and engineers began to arrive. However, as ZAFT had recommended, the returnees now traveled directly to Leipzig. Admission procedures, including career counseling, were carried out in a hotel there. Moreover, funds were provided for the salaries and the hotel accommodations of the specialists until employment contracts could be completed. As the importance of the repatriated experts grew, the number of persons involved in their supervision was reduced "for reasons of vigilance," and the party apparatus of the Socialist Unity Party of Germany (*Sozialistische Einheitspartei Deutschlands; SED*) directed greater attention to the integration processes.

Those returning from armaments-related industries had ideas about their future career plans that went beyond the sovereign powers of the GDR leadership. Thus, these specialists apparently pushed for the resumption of aircraft construction, but by the end of 1953 the Soviet side had not yet made a conclusive decision on this. Accordingly, the remigrants had to be given non concrete and unsatisfactory answers in response to their demands. Thus, the special welcoming ceremonies that some of the enterprises of long tradition, such as Zeiss of Jena, had organized for "their" specialists had — as one report commented — "especially depressing" effects on the others.[22] Moscow did not decide until May 1954 to recommence building aircraft in the GDR. Accordingly, the 307 specialists still in the Soviet Union were to return to the GDR at the beginning of June. The responsible authorities in the SED party apparatus received evaluations of the returning experts made by the Soviet side. At the same time in the USSR, the basic design for a new transport plane was still completed in conjunction with the specialists there under the direction of Brunolf Baade, the German chief design engineer. Since the Soviet dismantlement of East German industry left the GDR without its own research and test facilities, all work in these areas for the next two years was predicated upon Soviet support. This guaranteed the USSR continued and complete access to the technical knowledge and know-how of the German specialists.[23]

Though by the end of 1954, 91.4% of the specialists had been repatriated, those experts who were more prominent and had taken part in the development of weapon systems still had to remain in the USSR for a "cooling-off period."[24]

However, at this time, representatives of the SED party apparatus established contacts with the responsible Soviet state authorities in administration and security, inspected existing documents, and even spoke directly with selected specialists. They compiled comprehensive material that included professional assessments of the specialists, evaluations of their political attitudes and their contacts with the West, and operational data collected about them by state security. In conclusion, an assessment was given of the chances of getting them to commit themselves to the GDR and of their possible intentions of going to West Germany. The return of 139 specialists (or, including family members, of 405 persons) was planned for the end of 1954. Of these specialists, 104 were in Sukhumi, 26 in the Volga region, 5 in Moscow, 2 in Kharkov, and 1 each in Voronezh and in Rostov-on-Don. They included 1 professor of chemistry; 19 doctoral-degree holders in physics, 6 in chemistry, and 4 in medicine; 22 engineers or design engineers; 9 chemists; 2 physicists; 57 skilled laborers; 1 journalist; 1 student; and 17 specialists without a profession. Following detailed personal characterizations of Manfred von Ardenne (who was considered by the GDR side as the leading figure among the specialists) and of Nikolaus Riehl (the individual most highly commended by the Soviets), the material contained a list of further individuals who should be "operationally processed," i.e., put under surveillance by the Ministry of State Security. They allegedly had contacts with intelligence services or other "interesting associations with persons in capitalist countries," were former counterintelligence agents of the *Gestapo*, or had engaged in other "enemy" activities. However, there were also two specialists cited who were seen as having exhibited positive attitudes toward Soviet development. The Soviet side also furnished a list of those specialists who — due to the nature of the work they had performed in the USSR — it was especially interested in seeing stay in the GDR. In summary, however, the following conclusion was drawn: "The majority of doctors [i.e., holders of doctorates] and engineers will only make a decision regarding their work after their return to the GDR. Almost all intend to secure a good position. Their commitment to the GDR will be greatly influenced by the way in which they are employed. For this reason, it is also absolutely necessary to receive the specialists in a dignified manner."[25]

This group of specialists, which was mainly involved in the USSR in the building of the atomic bomb, came to the GDR in March 1955. The treatment of the team of Ardenne and his 15 specialists is the case best documented. The greatest possible efforts were made to implement the recommendation made in the preparatory material "to keep [him] in the GDR … by paying attention to his avarice and his need for admiration."[26] While still in the Soviet Union he purchased a piece of property in *Weißer Hirsch* in Dresden[27] and pushed forward the work of establishing his institute with funds from the Stalin Prize. Ardenne later reported that it had already received funding from Walter Ulbricht. In order to win the commitment of his staff of colleagues, he also had a cousin of his acquire properties for them in Dresden. Thus, upon his return on 25 March 1955, the entire institute was ready for operations.[28] The group was received in Frankfort on the Oder by the director of the department for technology of the Central Committee of the SED, Fritz Zeiler, and by the General Secretary of the Academy of Sciences. The latter was clearly only asked to appear for reasons of prestige, since it had already been decided that Ardenne would have nothing to do with the Academy.[29] Zeiler informed Ardenne of Ulbricht's desire to meet him just two days later. During the trip from Frankfurt on the Oder to Dresden, Ardenne confided to Zeiler "that he and his collective can create all necessary preconditions for the operation of an A[tomic]-pile, but cannot take on the actual task of building it." This message struck Zeiler as so confidential in nature that he did not even let the typist of the report know about it, only later writing in *A. Meiler* (A. pile) by hand.[30] It still remains unclear to what extent the decision to establish the Institute for Nuclear Research in Rossendorf near Dresden had already been made and whether this was connected to Ardennes's choice of Dresden.

This really did now account for the return of most of the specialists.[31] Nonetheless, some of those experts who had immediately expressed their desire to emigrate to the West had been detained for a much longer period of time. Apart from them, a number of individual scientists were apparently also still left in the Soviet Union. Thus, in 1956, attention was drawn to the cases of relatively well-known physicists still in the USSR. In March 1956, none of the returning scientists was found willing to go back to the Soviet Union for an additional two years (in order to work on the nuclear research institute of the Council for Mutual Economic Assistance, COMECON, that was to be jointly established in Dubna) until this question was cleared up. The physicists still in the USSR were Peter Adolf Thiessen, Heinz Pose, Robert Döbel, and Max Steenbeck.[32] The USSR then informed the GDR government that the four German nuclear specialists in question and two of their assistants were still in the Soviet Union. The following accounts were also furnished: Döbel had traveled to the GDR at the end of 1955, but had returned to the USSR after a period of time and requested appointment to the University of Voronezh. In the Soviet view, he could, if he wished be released from his obligations and return home. Pose worked at the United Institute for Nuclear Research. He could become the GDR representative at this institute if the GDR government agreed to this. Steenbeck worked at the Institute for Physics at the Ukrainian Academy of Sciences in Kiev. His departure for Jena, where his family lived, had already been set for August 1956. He wanted to wait until he arrived in Jena before making any future employment decisions. Thiessen was employed at the Moscow Physiochemical Institute of the Academy of Sciences of the USSR. His departure had been fixed for the end of 1957.[33] This Soviet information prompted the mission undertaken by Fritz Selbmann, who, as Deputy Minister President at this time, was responsible for the development of science and technology. In the spring of 1956, he held talks with the scientists in the Soviet Union in order to determine under what conditions they would return to the GDR.[34] He apparently also managed to allow some of them to return to the GDR earlier than scheduled.

In summary, it has to be said that the SED leadership sought from the outset to manage and monitor the integration of the specialists in keeping with their own priorities. Less attention was paid to the skilled workers, master craftsmen, and technical personnel who first returned than to the more prominent scientists and engineers. To some extent, this was the result of the improvisational manner in which integration initially ran its course until jurisdictional disputes and other uncertainties could be eliminated. But to a much greater degree, it resulted from general considerations of power politics which we will have more to say about below. In all, about 2,500 specialists had returned from the USSR by the end of 1958.[35] There is definitive evidence that 11% of these experts then went on to West Germany. Given the inexact nature of the card file in record to this question and the closing of the file after the return of the last specialists in 1958, it is fair to assume that 20%–25% of those repatriated in the GDR eventually migrated to the West. Of the remigrants, 45% were not university trained, i.e., they were skilled laborers, master craftsmen, and technical personnel. As their highest academic degree, 45.9% had a university degree in their field (das Diplom), 8.6% had a doctorate, and 0.5% had become professors. A comparison of the industries in which the specialists were employed prior to being taken to the Soviet Union and after their returned shows that some of the experts in aircraft and rocket construction moved to other fields, especially to other fields of mechanical engineering, electrical engineering, and precision engineering/optics. However, these moves also provided research institutes in the industry and the Academy of Sciences with new personnel.[36] This transfer of highly qualified employees into other branches of industry clearly provided these specialists with a source of employment. Whether it made sense in any other way cannot yet be conclusively determined.

The clearest qualitative effects of the return of the specialists can be seen in aircraft construction. It would not have been possible to reestablish this branch of industry after 1954 without these experts. And not just the more prominent scientists and executive engineers have to be taken into account in this context; for the mass of skilled laborers and master craftsmen were especially important. Moreover, this achievement also would not have been possible without Soviet assistance in the new construction and testing of airplanes, even if this support needs to viewed in a differentiated manner. Little is yet known about the influence exerted by the Soviet "style" of technological development on German natural scientists and engineers. The skilled laborers, master craftsmen, and technical personnel that returned in the first phase of remigration encountered a situation in the GDR predominated by minor corrective innovations aimed at removing bottlenecks in production. There was no great demand for more basic innovations of either the products or the production process. It is likely that the returnees were well-acquainted with the "improvising" that this situation called for, since in the Soviet Union (even though they had been provided with everything possible) they still had had to produce a large part of the machines, tools, and spare parts that they required themselves. Nevertheless, the importance of this "fit" should not be overestimated, since the specialists who had remained in eastern Germany had gained the same kind of experience dealing with the scarcity of the first postwar decade.

More intensive efforts were made on behalf of the later-arriving groups of more well-known scientists. Most of these scientists were initially impressed by the attention shown them by leading politicians, such as Ulbricht and Selbmann.[37] In my judgment, the interest of the latter was genuine and not feigned. For in the mid-fifties, the party leadership sought to get industry to make greater use of advances in scientific technology. Thus, they considered the returning specialists from the Soviet Union not merely as a scientific enrichment of their labor force, but also as a source of recommendations and advice for directing and organizing research. The GDR leadership may also have prided itself with having indirectly done "its" share (by way of these specialists) in maintaining the superpower status of the Soviet Union.

On the other hand, the GDR leadership also acted in terms of its own interests in continued power: they knew that such attentions would flatter the vanity of such scientists and help in winning their commitment to the GDR. This was part of broader calculations that involved competition with the Federal Republic of Germany: especially in view of the fact that most of the top scientists returning had good chances of finding interesting career opportunities in both East and West. Success in getting these prominent scientists and engineers to commit themselves to the GDR provided the country with urgently needed prestige, and, especially in the eyes of these specialists' peers, with a certain degree of legitimation.

The impact of specialists well-known even prior to 1945 on young scientific talent was not limited to their research projects. A large part of the following generation of natural scientists and engineers did not remain unaffected by the German scientific traditions these specialists embodied and their readiness to commit themselves to the societal goals and institutions (the Research Council, Peace Council, and similar institutions) they considered beneficial. It should also be noted that the more prominent of the scientists and engineers returning from the Soviet Union played a crucial role in the successful introduction of planning into the field of research. Without them and the trust they enjoyed in the Soviet Union, the Research Council (*Forschungsrat*; founded in 1957) never could have had the impact it did, especially in its first operational phase, prior to the building of the Wall. By passing on their experiences with major research projects and the research planning it entailed, these specialists made it possible to establish agencies of this kind, which were so markedly conceived in Soviet terms. Given that the basic conditions in the GDR required the concentration of all available resources, it certainly made sense to establish such an advisory institution as the Research Council, if its autonomy could be respected. This autonomy was, however, limited from the outset. Since all of this, however, involves issues that are very difficult to quantify and to verify, further research is called for here.

NOTES

1 Manfred von Ardenne, *Sechzig Jahre Forschung und Fortschritt. Autobiographie* (Berlin: Verlag der Nation 1988); Ferdinand Brandner, *Ein Leben zwischen den Fronten. Ingenieur im Schußfeld der Weltpolitik* (Munich, Wels, 1976[2]); Nikolaus Riehl, *Zehn Jahre im goldenen Käfig. Erlebnisse beim Aufbau der sowjetischen Uran-Industrie* (Stuttgart: Riederer 1988); Max Steenbeck, *Impulse und Wirkungen, Schritte auf meinem Lebensweg* (Berlin: Verlag der Nation 1978); Peter Adolf Thiessen, *Erfahrungen, Erkenntnisse, Folgerungen* (Berlin/East: Akademie-Verlag 1979).

2 Ulrich Albrecht, Andreas Heinemann-Grüder, Arend Wellmann, *Die Speziasten. Deutsche Naturwissenschaftler und Techniker in der Sowjetunion nach 1945* (Berlin: Dietz 1992).

3 The precise analysis of the card file to which recourse is made here now available: Burghard Ciesla "Der Spezialistentransfer in die UdSSR und seine Wirkungen in der SBZ und DDR (1945–1958)" *Aus Politik und Zeitgeschichte* (1993) *B49–50/93*, 23–31. In terms of the absolute number of the returnees in individual years, minimal differences (which in individual cases cannot be cleared up) exist between the totals reach by Ciesla and those presented here on the basis of contemporaneous documents. The delayed recording of this data into the card file may account for these differences.

4 MdI, Berichte zu den Übernahmen der Sondertransporte mit deutschen Spezialisten aus der UdSSR von September/Oktober 1950, in: Stiftung Archiv der Parteien und Massenorganisationen der DDR im Bundesarchiv. SED holdings (hereafter as SAPMO, SED) IV 2/13/389.

5 According to present information, they were probably employed in the development of electronic instruments, and in television and high-frequency technology. Cf. Albrecht "Spezialisten" p. 163.

6 MdI, Bericht über zurückgekehrte Spezialisten im Jahre 1950, 25.10.51 in: SAPMO, SED IV 2/13/389.

7 MdI an das ZK der SED, Sekretariat, 22.2.51: Politische Betreuung deutscher Spezialisten, in: *Ibid.*

8 Zentralamt für Forschung und Technik (ZAFT), W. Lange an Heinrich Rau, 21.11.50, in: Bundesarchiv, Abteilungen Potsdam (hereafter as BAP) DF4/463.

9 Ministerium für Arbeit und Gesundheitswesen Sachsen-Anhalt, Becker an Ministerium für Arbeit und Gesundheitswesen der DDR, Minister Steidle 27.10.50: Zurückgekehrte Spezialisten aus der UdSSR, in: BAP DQ2/1789.

10 Hauptabt. Arbeit: Vermerk, 30.10.50: BAP DQ2/1789.

11 Calculated according to BAP DE1/28332.

12 MdI an das ZK der SED, 26.10.51: Bericht über zurückgekehrte Spezialisten im Jahre 1950, 25.10.51, in: SAPMO, SED IV 2/13/389.

13 Calculated according to BAP DE1/28332.

14 The relevant correspondence is found in BAP DF4/460.
15 Ciesla "Spezialistentransfer", 28.
16 MdI, Bericht über zurückgekehrte Spezialisten im Jahre 1951, 8.10.51, in: SAPMO, SED IV 2/13/389.
17 Lange an Rau, 21.11.50.
18 ZAFT, Lange an MdI, Büttner: Berufsberatung von Spezialisten aus der UdSSR, 30.11.50; MdI an ZAFT: Berufsberatung von Spezialisten aus der Sowjet-Union, 8.12.50, in: BAP DF4/463.
19 Ministerium für Arbeit, Aktenvermerk: Spezialisten aus der SU — Fachliche Betreuung, 22.12.51; Bericht: Zurückgekehrte Spezialisten aus der SU, 24.12.51, in: BAP DQ2/717; Abt. Arbeitskräftelenkung, Bericht über die Betreuung der aus der UdSSR zurückgekehrten deutschen Spezialisten, in: BAP DQ2/1789; ZAFT, Hausmitteilung: Meyer an Lange, 12.1.52, in: BAP DF4/463.
20 ZAFT, Lange an SPK, Kerber: Betreuung der Spezialisten, die aus der Sowjetunion zurückkehren, 18.7.52, in: SAPMO, SED IV 2/13/389; Betreuung heimkehrender Spezialisten aus der SU, 28.6.52, in: BAP DF4/463; Ministerium für Arbeit, Abt. Arbeitskräftelenkung an Hauptabt. Arbeit: Spezialisten, 4.11.52, in : BAP DQ2/1789.
21 Ciesla "Spezialistentransfer", 28.
22 MdI, Bericht über 5 Transporte mit deutschen Spezialisten aus der UdSSR. Abt. Maschinenbau an Ziller: Rückkehr von Spezialisten aus der Sowjet-Union, 25.1.54, in: SAPMO, SED IV 2/13/389.
23 Fritz Zeiler: Bericht über die Besprechung anläßlich des Besuches vom 18.–21. Mai 1954, 24.5.54, in: SAPMO, SED J IV 2/202/56.
24 Ciesla "Spezialistentransfer", 28.
25 Bericht über die zurückkehrenden SU-Spezialisten, 31.12.54, in: SAPMO, SED J IV 2/202/56.
26 *Ibid.*
27 Traditionally, a residential area of the well-off in Dresden.
28 Vortrag von Prof. von Ardenne in der Kongreßhalle Berlin, 17.3.70, in: SAPMO, SED IV A2/2021/367; Ardenne "Sechzig Jahre" p. 250, 255. Cf. Albrecht "Spezialisten" p. 73.
29 Cf. Ardenne "Sechzig Jahre" p. 264. Ardenne does not mention Zeiler, a fact that probably can be explained in terms of his reluctance to admit the central influence of the party apparatus.
30 Abteilung Technik, Rückkehr der deutschen Spezialisten aus der Sowjetunion, 25.3.55, in: SAPMO, SED J IV 2/202/56.
31 Albrecht "Spezialisten" p. 182.
32 Ernst Wolf: Kurze Notiz über die Moskauer Verhandlungen zur Gründung Vereinigten Instituts für Kernforschung, 28.3.56, in: SAPMO, SED J IV 2/202/28.
33 (Memorandum about individual specialists, Russian), in: SAPMO, SED NL90/472.

34 Steenbeck "Impulse" p. 318.

35 According to Ciesla, the card file of the MdI contains 2,401 specialists. Due to various uncertainty factors, Ciesla, however, corrects this number to about 2,500. Cf. Ciesla, "Spezialistentransfer" p. Thus, the figure of 2,370 natural scientists and technical experts that Albrecht *et al.* consider verified was pretty close. Their estimation of a total of 3,000 specialists, though, was too high. Cf. Albrecht "Spezialisten" p. 178, 182.

36 Ciesla, "Spezialistentransfer", 29.

37 Ardenne "Sechzig Jahre", 271, Steenbeck "Impulse", 339.

CHAPTER 10

The Politics of Ambiguity: Reparations, Business Relations, Denazification and the Allied Transfer of Technology

Johannes Bähr, Paul Erker and Geoffrey Giles

The importance of science and technology in World War II profoundly affected the Allies' view on how to impede the resurgence of a defeated Germany. As John Gimbel's term "intellectual reparation" implies,[1] the traditional "victors-and-vanquished-concept" could no longer work. The Allies were not only interested in controlling and limiting the enemy's economy, but also its scientific and technological infrastructure. In addition, they wanted to make use of Germany's achievements in the production of knowledge. The latter had even to include the hiring of German scientists, who were responsible for Germany's technological progress, but were often tainted with their Nazi past.

In 1945 the Allies experienced quite extraordinarily complex problems in determining who was a Nazi. The initial difficulty was a result of their not deciding in advance of the cessation of hostilities how precisely to define this term. That none of the several drafts of the occupation handbook produced by 1944 by SHAEF did so was a reflection of Roosevelt's unwillingness to plan detailed occupation policy until after the defeat of Nazi Germany.[2] What has emerged from the recent literature on the reconstruction of the German universities after the Second World War is the makeshift nature of policy-making in all four zones.

Perhaps it should therefore not be surprising that at the various symposia held in the last few years, reuniting former university officers of the four occupation zones, there was some confusion over the exact meaning of the term "denazification" among them, too. Did it refer to the simple dismissal of tainted faculty, or was it rather the lengthy process of reeducation within the German universities that emerged as a goal of the Allies? Those dismissals were indeed extensive in German universities after 1945, but we may legitimately ask whether they were comparable in their damaging effects to the Jewish purges of 1933/34?

The best-known and most obvious example of such intellectual plunder is in the area of atomic physics. The ALSOS Mission moved swiftly, apprehending the last of Germany's top ten atomic scientists in Hamburg on 3 May 1945, and spiriting them all away to England two months later. In October the Mission, comprising a team of over 100 people, was disbanded.[3]

The transfer of technological expertise did not always entail the physical removal of scholars from German soil. Especially in the Soviet Zone, scientists often had to carry out contract research for the Soviets. The Mining Academy at Freiberg provides an example of policy confusion in that zone as well. A mere three weeks after Soviet troops had occupied the town, a provisional senate of college dismissed all the most active Nazi faculty members. Almost immediately the Soviet occupation forces decreed that everyone, regardless of their political record, would be granted the possibility of staying on, in order to work on Soviet research projects. Then in November 1945, responding to criticism from SMAD that purging of the teaching staff had been unsatisfactory, the Department of Education of the State of Saxony ordered all former members of the Nazi Party and all its affiliated organizations (down to the NSKK[4]) to be sacked. The dismissals at the Mining Academy totaled sixty, a purge so extensive that the success of Soviet research projects was obviously threatened. Realizing this, the Soviet overseers of the Academy interceded with the Supreme Command of the Soviet Armed Forces to achieve on 8 December a rescission of the order, reinstating all the dropped faculty, if not necessarily to teach, at least to continue with their research.[5]

Here, then, is an example of the widely observed phenomenon in the Soviet Zone of sweeping dismissal and creeping reinstatement. It suggests a triumph of continuity rather than change, and a failure of denazification.

Did the latter mean the same thing to all the Allies? Certainly there was an unstated agreement in the separate policies of the Allies toward institutions of higher education in the weeks following the end of the war. Yet it suggested that denazification was not the first priority. As James Tent puts it: "All four Allies exploited the first phase between occupation, unconditional surrender, and the rulings at the Potsdam Conference (17 July to 2 August 1945), in order to obtain the maximum amount of plunder from among the universities, and especially the research institutions. "The fact that an immediate reconstruction of higher education was not uppermost in the minds of, for example, the American government, is indicated by the instruction, in the crucial decree JCS 1067, which came into effect in April 1945, that initially all research activities should cease, and all laboratories should close, until such time (and this remained vague) that "careful investigation" should establish that the research would not contribute to Germany's future war potential.[6]

And yet dismissals did very quickly occur, surely an indication to the Germans, if we wish to consider their "meaning," that the institutions were being prepared for reopening, and the positions would be filled again by ideologically untainted personnel. But this was not denazification. There was indeed some confusion at the time, which led to a French clarification in April 1946, to the effect that: "La dénazification [est] distincte de l'épuration."[7]. The memorandum made it clear that l'épuration or Säuberung was something imposed upon the Germans by the victors, whereas denazification was an activity that only the Germans themselves could fully achieve, and embraced not appointments, but a whole change of attitude, which would become visible through, for example, the rewriting of text books. If the Germans were evidently unclear about this, the fault lay partly with the occupation forces, who were not always aware of the distinction themselves.

In some regions, denazification was understood as a simple manoeuvre of turning the clock back to Weimar, or even further. The maddeningly effective deviousness of Alois Hundhammer, the Bavarian Education Minister, in circumventing American attempts to democratize the school system, is perhaps the best-known example of this.[8] The British did intervene directly on occasion in the educational sphere, as when they banned model airplanes from their zone altogether, mindful of the role these had played in the 1930s in promoting airmindedness and thus representing a contributory factor toward militarism.[9] But on the whole their approach was much more one of laissez-faire, and Germans, especially professors, were treated as if they possessed sufficient responsibility to make the right decisions. Some British politicians were not so sure about this, and the prominent Socialist, Ernest Bevin, noted as late as November 1947 in a speech to the heads of British universities that their influence on their German colleagues was paramount because "the Germans did not know where they were going and there was in fact a spiritual and moral vacuum in the country. There was a danger that this vacuum might be filled by Communist materialism and it was essential to prevent this by making sure that the cultural standards of the West prevailed."[10]

A does of anti-communism was needed, then, in order to ensure that denazification did not go too far in the opposite direction.

In the American Zone there existed what appeared to be competing agencies in the reshaping of university life. The Counter-Intelligence Corps (CIC) took care of Säuberung primarily, while the Education and Religious Affairs Branch of the military government handled denazification in the sense of rebuilding the university. Yet they constantly trod on each other's toes, as has become even more evident at conferences held in the last two years at Tübingen and Heidelberg.[11] The CIC viewed its role very much in terms of a policeman: the German universities were full of villains and must be purged to the fullest extent. This policy had repercussions even outside the US Zone. L.H. Sutton, one of the very first University Officers, assessed the situation at Göttingen in January 1946 in the following terms: "The original object of denazification was, in education, to eradicate Nazi teachers, Nazi books, and everything contaminated with Nazi ideas. The present object is to be able to show, for the benefit of our Allies, a number of dismissals in the British Zone which is at least as great as the number for the Allied zones."[12]

In other words, a serious question of ideological reform had degenerated in a simple competition of numbers. Immensely complex as that problem was, there had been little thought given to its solution among the occupation staff in any of the zones, apart perhaps from that of the Soviets. With the notable exception of Edward Hartshorne, the towering giant of educational reform in the US Zone before his mysterious murder in August 1946, most members of E & RA arrived in Germany in woeful ignorance of the German university system, though some of them were faculty in American universities. Any specific preparation for their task as university control officers was virtually non-existent, and certainly not on an ideological level. It was cheerfully assumed that individuals coming from a democratic society would know democracy when they saw it and be able effectively to encourage its growth in defeated Germany. John Gimbel reflected: "I wonder how much theoretical discussion there was about the nature of fascism or totalitarianism. I am not sure that we read for example Franz Neumann's Behemoth, which was published in 19[44]. We got lectures from our professors and I remember that they gave us a sort of leftist interpretation of fascism."[13]

Yet here lay the real problem of occupation: the ideological reorientation of a nation. This was a task, unless we are to believe that Germany was to be abandoned to the sort of miserable status wished upon it by Henry Morgenthau, that demanded painstaking preparation, acute insight and sensitive execution. The harvesting of intellectual reparations was relatively easy: gather up the scientists and scholars, many of them alarmed at the prospect of communist control anyway, and ship them off to a better life in Britain and the US. A straightforward bureaucratic exercise — what could be simpler? Dismissing politically suspect faculty was also uncomplicated — if one acted swiftly, without probing too deeply into the details of each individual case. But these were not constructive policies for the future of Germany. The country had not vanished from the map. That problem remained, and in my view as the most central one facing the Allies. There are many signs that they acted as though they did not realize this. In the worlds of one British official which might stand for all the Western Allies: "It was on the positive side that our thought was much less definite and much less distinctly formulated."[14]

Significantly, it was this problem which tripped up the Nazis themselves in the universities after 1933.[15] Purging the old is an easy task, filling the void created is an altogether different and infinitely more difficult task. Nazi scholars found it hard to explain precisely what Nazi science was. It was easier to pontificate on what it was not. And thus those other, non-Nazi scholars who collaborated to some degree with the new regime were able to ensure that their influence was not altogether lost, as they strove to maintain the continuity of the institution, and with it their own livelihood. After 1945, a parallel situation existed: new ideological masters, faculty purges, an exodus of scholars to other countries. James Tent sums up the question in carefully measured terms:

"Denazification while imperfect did manage to eliminate enough Nazis long enough from positions of influence to allow the new society, and in this instance the universities, to begin functioning again under the leadership of largely uncompromised individuals." And then he adds: "While criticized at the time and afterwards as being either unfair or inadequate, denazification never descended to the level of a mockery of justice as happened with Allied efforts to try war criminals in Weimar Germany after World War I."[16] We tend to agree with Mark Walker that we are still too close to the collapse of the GDR to make such an interpretation, such a positive comment, about the 'destasification' of universities in the New German States which is still occurring today.

The Allied transfer of technology from Germany after World War II took place under conditions without parallel in the history of the leading industrial countries. As a result of the unconditional surrender of Germany, the total military occupation of the country, and the setting up of military administrations in the zones of occupation, the victorious powers were in a position to appropriate the intellectual property of firms and private individuals either free of charge or at reparation prices, and to transfer whole teams of scientists and engineers.

The original objective of the Allies aimed to a decisive reduction in the economic-industrial potential of the vanquished aggressor. Moreover, the Allies added the endeavour to build up their own research and development capacities and make up for technological deficits in various fields.

The rivalry among the victorious powers gave further significance to the transfer of technology from Germany. Immediately after the war a race between the superpowers began, anticipating the constellation of the Cold War. The American Project Paperclip and the Soviet Operation Ossavakim of 1946 intended to remove technological know-how from the grasp of the other superpower.

The reparations' balances do not contain "intellectual reparations". First, there are statistical reasons not to include the confiscation of intellectual property into the reparation account. This is, because one can only speculate on the value of this transfer for both the Germans and the Allies.

German estimates of the value of the Allied transfer of technology vary between 4.8 and 12 billion US-$.17 Christoph Buchheim gives a value of 2.8 billion US-$ by assigning to it 65% of the expenditure for industrial development. The latter proceeds from the assumption that a major part of its research and development expenditure of German industrial in the field of armaments technology after the end of the war was a misinvestment.[18]

In the Allied balance, these countries' own development costs must be taken into account, the evident differences between the Western Allies and the Soviet Union not being able to be established.

Second, from the viewpoint of economic and technological history, the application of the concept of reparations to the field of research and development is dubious, for reasons of methodology.

The transfer of technological knowledge took place within a totally different frame of reference from that of goods and equipment. Know-how is tied to highly-qualified specialist personnel and integrated into long-term processes of research and development. The transfer only means a reduction of new knowledge in the country of origin when those in possession of it were permanently lost to the country.

Therefore, the effects of the transfer differ fundamentally from those of material reparations payments. Despite the considerable extent of the Allied transfer of technology, it did not destroy the substance of the original potential of Germany. In the leading fields of West German industry, new knowledge was not lost by the transfer, since the specialists were still present. As far as the firms were concerned, all that happened was an increase in the pressure of competition. Exceptions to this were areas particularly relevant to armaments, such as missile research, where a majority of the experts left Germany. However, these areas would hardly have been able to expand here after the war.

On the other hand, West Germany's industry was also able to profit from integration into the Western sphere of influence. Contacts between West Germany and the victorious Western powers were not limited to the activities of the American, British or French authorities. On the level of the firms, traditional contacts were revived soon after the end of the war. Little is known about the forms of this cooperation in the early postwar years, but it may be assumed that firms that were interested in the restoration of longer-term business relations banked on an exchange in the technological field.[19]

The transfer of technology to the Soviet Union largely took place via newly founded Soviet experimental and construction centers in East Germany, at which German experts from numerous fields worked. Some of these centers expanded into major plants which had an influence on the later development of East German industry — such as the Oberspreewerk in the eastern sector of Berlin. [20]

It was only with Operation Ossavakim in the autumn of 1946, the deportation of over 2000 scientists, engineers, technicians and skilled workers, that this specialist's potential was transferred to the Soviet Union.[21]

The German experts returned from the Soviet Union in the early 1950s. The majority of them remained in the GDR, and had a lasting influence on industrial research and research policy in the GDR. The fact that East Germany later fell behind in the technological field was not due to the transfer to the Soviet Union but to the barriers within the GDR economy caused by the system.

The transfer of technology to the Soviet Union was equally able to form the basis of new contacts that were favourable for the position of the GDR within COMECON. However, the German scientists were not integrated into Soviet research. In contrast to most of the experts who transferred to the USA, they were sent back from the Soviet Union as soon as the technological gap there had been made up.

The success of a transfer of new knowledge by means of personnel depends on the readiness to cooperate of those concerned. The Allied transfer of technology was certainly attractive to German experts. It opened up to them career opportunities and material security on which they were not able to count in postwar Germany. Politically tainted persons were also able in this way to evade the denazification regulations.

It is evident that a brain drain to the USA would have taken place after World War II even without the intervention of the American authorities. The transfer would have taken the form of a migration of labour comparable to the present emigration of atomic scientists from the CIS. The losses to the German economy would hardly have been less in this case.

Most of the German specialists who worked for Soviet authorities after the war were also acting in their own interest. The Soviet experimental and construction centers in East Germany attracted able scientists and engineers. According to a report of the American secret service, these also included experts who were accounted politically suspect in their firms.[22] It was only in the course of Operation Ossavakim in October 1946 that measures of force were applied to the majority of the specialists working for the Soviet Union.

Even under the particular political conditions obtaining, the transfer of technology from Germany after World War II was marked by typical elements of the passing on of new knowledge that contrasted with the mechanism of reparations. The importance of the transfer of technology in the Allies' reparations policy, which Gimbel has made evident with the notion of "intellectual reparations", is only one factor in a complex process which also contained elements of a labour migration and incentives for German experts.

The Allied technological transfer of 1945/46 is only a small part and a special case of the economic, scientific, and technological relationship between Germany and the United States that has existed for a long time and is much more complex.

The prewar (and that means until 1942) business relations, for example, played a role in this relationship, which continued right after the end of the war despite the forced transfers ordered by the Military Government. At the same time, a pattern emerged of the counterflow of technological and scientific information from the United States to Germany.

In addition to that, as far as the effects are concerned, we have to distinguish sharply between technological transfer within the context of the Cold War and, as such, under the aspect of military security — or within the context of foreign markets and, as such, under the aspect of economic competition.[23]

At first glance, the integration of the German experts — be it their know-how, their style of research work, or their relationship to industry — seemed to have been successful. But there are some strong indications that cast some doubt on the clarity of this picture, doubt regarding the state of the art of German aerodynamics, doubt on the alleged backwardness of American aviation science, and doubt concerning the story of the successful integration of German scientists and technology.

First, Leslie Simon's and, at the very least, Helmuth Trischler's books have already shown that around 1943–44 German aviation research lost its theoretical and experimental advantage, operating without coordination and wasting both time and money.[24] German aviation research since 1943 comprised little else but unsystematic experimental work, a steady trial and error game.

When the Military Government in Germany imposed its research restrictions, the remaining scientists regarded the destruction and dismantling of the wind tunnels as a stroke of good luck ("geradezu als einen Glücksfall"),[25] because it forced them to return to doing the basic research that had been suppressed in Nazi Germany.

Concerning the integration of the German scientists, we need to know more of the details before we can form a final judgment: How was German scientific knowledge introduced into the US Air Force labs? How much access were German scientists given to Air Force know-how? They came as enemies and nobody knew whether and when they would leave to become future competitors.

Ultimately and without much doubt, the Air Force profited from the work of German scientists, but much less than Ciesla's essay suggests. In addition, it remains problematic that the discussion of "Paperclip" concentrates too much on "big science" fields like aeronautics, aviation, and nuclear physics. Moreover, the story of aerodynamics is a special case of postwar technological transfer.

An example of the "normal case" are the chemical technologies, analyzed in Stokes' essay. Here the evidence appears to indicate that the mutual repercussions, that is, the losses for German companies as well as the gains for American firms, were rather small.

Above all we must take into consideration two decisive factors that modify the importance of the forced Allied technology transfer after 1945.

First there was an "incomplete understanding of science and technology, a misconception as product rather than process on both the victorious Allies and the vanquished Germans" (Stokes). The Allies thought that science and technology could be channelled, controlled, taken away here, and implemented in their own countries. Both sides soon realized that this would not work.

A second point related to the first in a certain way, has not been mentioned in the two essays: There is a tradition of technology transfer between the late 1930s and early 1960s, especially if we look at relations between German and American firms.

Close relations through cartels, patent and licenses exchange agreements, and arrangements concerning sales and manufacturing existed throughout the interwar period, and was only interrupted for a short time and restored as soon as it was possible. It was not at the end of the 1940s that private companies and the public complained about the brain drain but in the mid-1950s and early 1960s, when the catch phrases of "buying out Germany," "decline of German science," and "loss of technological competence" made the rounds. Indeed, this exodus of German scientists and engineers to the United States was much more important than that of 1946. Therefore Stokes's plea "to shift the focus of investigation from Allied economic gain (or German loss) to the direct technological and scientific impact of the seized or exchanged information or personnel" can be supported. The problem of reparations is not the really exciting and interesting theme, but rather the mechanisms of technology transfer under changing conditions, and forced Allied technology transfer is only a special case in this light.

This argument can be well illustrated by another example, the-one of the electrical industry.[26] Four points have to be considered.

1. The FIAT reports and the captured know-how proved to contain information only of minor value to American companies. In another article, Stokes already emphasized this point for the chemical industry: "Often only the Germans possessed the technological know-how to make the knowledge productive."[27] In this connection, we know nothing about how successful German firms were in hiding capital, know-how, and patents abroad just before the war ended. At least one of the leitmotivs for the "Safehaven" project of the American Military Government was to reveal hidden capital.

We have to add that already in the late 1940s West German companies possessed enough valuable bargaining chips to start negotiations with Anglo-American firms about know-how and patent exchange. The semiconductor technology developed by Siemens is a good example. Germany's biggest electrical company demonstrated that forced Allied technology losses did not prevent it from initiating a successful strategy to regain quickly its global market position.

From the beginning, Siemens laboratories concentrated on rectifier technology, which had been large neglected by American electrical firms in favour of the new transistor technology. As a result, after five years of research and development the Siemens thyristors prevailed on the international market (ca. 1955).

On the whole, German producer and capital goods industries as traditional main export lines on the world market were not inferior to those of the competitors, as far as the technology was concerned.

Above all, two factors — quality and terms of delivery — frequently turned out to favour of German industries. Thus, losses in some of the leading technologies owing to Allied restrictions, on the whole, were easily compensated for. The competitive edge given to Allied firms quickly disappeared. German firms reconquered the foreign markets with old and proven technology but at the same time carefully prepared the groundwork for the implementation of new "American" technologies.[28]

2. The FIAT reports and the know-how transfer were also of minor importance because the former US competitors and business partners were already aware of technological approaches in Germany. For example, IG Farben maintained its close exchange relationship with DuPont until 1942.

Siemens did likewise with General Electric, Westinghouse, and International Telephone and Telegraph (ITT). Indeed, in the filed of long distance telecommunications (Telod) or in the field of telephone apparatus (Telap) — as well as in many other sectors of research and production — long term agreements had been made with the most important American electrical firms. These agreements stipulated the exchange of patent rights and technical information or gave Siemens "exclusive rights to use Westinghouse inventions" and vice versa.

This cooperation continued even during the war. When the United States entered the war in 1941, Siemens' Director Lueschen, for example, met with representatives of the ITT board of management in Switzerland. At this meeting both parties agreed to continue to cooperate as long as possible. They also agreed to "avoid everything as far as possible that would make difficult the conclusion of new agreements after the war" as well as "to avoid competition in the registration of new patent rights during the war."[29]

Obviously, this is one of the reasons why there was almost no transfer of scientific personnel from Germany's chemical and electrical industries to the United States. Allied officers even failed to recognize the value of Germany's achievement in the development of semiconductor technologies. When the officers scrutinized the industrial research and development (R&D) of Germany's electrical industry in 1945, they found "a considerable amount of theoretical and experimental work on semiconductors, especially at the Siemens and Osram laboratories, but no striking development in these fields."[30]

3. The international scientific community experienced a quick reconstruction after the war, which led to the emergence of new avenues of technological and scientific exchange. As a result, important areas of scientific knowledge experienced rapid change.

These developments helped firms like Siemens to succeed in rebuilding quickly its R&D capabilities. Certainly, the domain of knowledge in nuclear physic at Siemens, for instance, had been weakened by the transfer of its research head, Gustav Hertz, and five of his most able colleagues to the Soviet Union. On the other hand, the return of Prof. Finkelnburg, a former Siemens scientist who had been working in the American electrical industry in 1949, granted Siemens a better and brand new source of know-how in nuclear physics. Also in the field of semiconductor physics, Siemens could take advantage of the return of German scientists who previously had worked in the US. For example, the former manager of Siemens & Halske's central laboratory, Hans F. Mayer, returned from the US to Germany in mid-1948 after fulfilling his obligations as a guest professor there. Now he could inform his colleagues at Siemens about recent technological developments in America.

Restrictions on research imposed by the Military Government even offered the opportunity to German industry to make up for what had been missed in the last years of the war when basic research had been neglected in Germany. One field where the return to basic research proved invaluable was the peaceful use of nuclear energy. As a result, only the later West Germany was able to apply it in the near future.

Here, too, the ban on applied research was not serious, since the question remained unanswered as to where, in fact, the line would have to be drawn between the basic research that was permitted and the applied research that remained banned. At least, the replacement of destroyed or dismantled R&D units offered the chance for German companies to select the scientific personnel carefully and to deflate the bloated personnel structure of the war period.

Whereas in 1948–49 Siemens started with a small, flexible research organization, General Electric entered the international research competition with a rather rigid and bureaucratized organizational structure in the 1950s. This bureaucratic structure included also a greater staff of personnel. The better efficiency of Germany's research and development contributed to the fact that German industrial research could recover lost ground within a few years.

According to these arguments, it is evident that the one-way flow of knowledge and information out of Germany lasted only a short time and was quickly replaced by mutual exchange of technologies. Regarding the case of the chemical industry one can add that the replacement of coal tar chemistry as the major technology applied by oil chemistry led to the expiration or devaluation of many patents.

4. Finally, one has to put the well-known "not-invented-here-effect" into consideration. This effect imposed limits on the reception of the captured know-how by American industry. Again, the electrical industry provides an example. Several American big electrical companies like RCA, General Electric, or Westinghouse exhibited no interest in the magnetic tape recording technology, developed by AEG/Telefunken and ready for production as early as the mid-1930s. When the FIAT officers brought the highly advanced tape technology in the United States in 1946, companies like RCA preferred to invest in their own technological concepts rather than in AEG's research results.

In contrast to bigger companies a rather small company — Ampex — founded in 1944, succeeded with AEG's tape recording technology.[31] Even though Ampex experienced a commercial success in applying the technology, the Ampex example is rather the exception than the rule.

Reviewing the existing case studies, including the ones presented in this volume, it becomes apparent that the main and foremost intention of American companies was to examine German industries closely rather than to exploit and plunder it. American companies wanted to get a first-hand impression about the possibilities and chances for gainful technological and commercial cooperation and capital investment in Germany itself. Volker Berghahn and Paul Friedrich proved this assumption in a recent study. They chose the rubber industry as a good example. American technical officers, who were at the same time managers of the US rubber industry, after their investigations into Germany's rubber industry quickly decided to transfer American technological know-how to Germany. By investing a good deal of money, the American rubber industry then intensified its business relationships in Germany itself.[32]

All in all, indeed, the forced technology transfer promoted the integration of the then emerging East and West German industries into respective spheres of influence dominated by one of the two superpowers. Rather this integration than the technology transfer itself promoted the growing division of German technological culture in the years following the occupation period.

In the Western zones, the technical investigations and transfers at least established and reestablished professional contacts between German and American firms. As Gimbel has written, FIAT functioned "as a conveyer-belt for future business connections."[33]

This is the reason why the central question guiding historical research still remains: Why, in view of the postwar technological plunder, West German firms regained their strong world market position so quickly, be it in the commercial or in the technological field?

Much more than the West German case, the case of East Germany leaves a lot of questions unanswered. We need much more detailed analysis of the mechanisms of technology transfer to the Soviet Union and the repercussions this had on the remaining German firms.

Therefore, the prevailing major question for historical research on the East German case is different from the West German one: Was the forced technology transfer of the late 1940s responsible for the unfavourable conditions for East German industrial development in the early 1950s?

Nevertheless, it seems to be dubious that this question is really guiding us in the right direction of historical research. It is obvious that neither reparations nor the dismantling and plundering hindered East Germany's industries to a large degree.

It is preferable to say that other reasons led to East Germany's later decline. These reasons are the isolation of East German scientists from the international scientific community, the GDR's forced integration into the Soviet political and economic system, the specific organization of know-how in the GDR, and the inability of the centralized state planning system to generate technological innovations.

The latter leads to questions for future historical research that place the Allied technological transfer in a broader context: What is still lacking is the social and economic history of the Cold War. That should be the new perspective and the new framework that aims at an understanding of the historical phenomenon of Allied technology transfer from Germany after 1945.

NOTES

1 John Gimbel, Science, Technology, and Reparations. Exploitation and Plunder in Postwar Germany (Stanford CA: University Press, Stanford 1990), thereafter: Gimbel, "Science."

2 James F. Tent, "Denazification of Higher Education in US. Occupied Germany, 1945–1949," (thereafter: Tent, "Denazification"), in Manfred Heinemann (ed.), Hochschuloffiziere und Wiederaufbau des Hochschulwesens in Westdeutschland 1945–1952, Teil 2: Die US-Zone (Hildesheim: Verlag August Lax, 1990), 9. Thereafter: Heinemann, "Hochschuloffiziere 2."

3 Manfred Heinemann, "1945: Universitäten aus britischer Sicht," in Manfred Heinemann (ed.), Hochschuloffiziere und Wiederaufbau des Hochschulwesens in Westdeutschland 1945–1952, Teil 1: Die britische

Zone (Hildesheim: Verlag August Lax, 1990), 46. Thereafter: Heinemann, "Hochschuloffiziere 1."

4 Abbreviation for *Nationalsozialistisches Kraftfahrer-Korps* (National Socialist Driver's Association).

5 See text Giles.

6 Tent, "Denazification" 46.

7 Report of 8 April 1946 read out in discussion at the colloquium with former university officers of the French Zone, September 1985, in Manfred Heinemann (ed.), Hochschuloffiziere und Wiederaufbau des Hochschulwesens in Westdeutschland 1945–1952, Teil 3: Die Französische Zone (Hildesheim: Verlag August Lax, 1991), 41. Thereafter: Heinemann, "Hochschuloffiziere 3."

8 James Tent, Mission on the Rhine: Reeducation and Denazification in American-Occupied Germany (Chicago: University of Chicago Press, 1982), pages 133–135 and 141–152.

9 Heinemann, "Hochschuloffiziere 3." 1980.

10 Minutes of the 28 November 1947 meeting of the Committee of Vice-Chancellors and Principals, quoted in Heinemann, "Hochschuloffiziere 1," 21.

11 See for example Geoffrey J. Giles, "Reeducation at Heidelberg University," in press for Jürgen Heideking and Jürgen Herbst (eds.), Mutual Influences on Education: Germany and the United States in the Twentieth Century; and also Geoffrey J. Giles, commentary at the Heidelberg symposium, to be published in Hartmut Lehmann (ed.), Heidelberg 1945/46: Amerikanische Besatzung, Universität und Stadt Heidelberg und die Heidelberger Beiträge zur Erneuerung des politischen und kulturellen Lebens in Deutschland nach der Katastrophe des Nationalsozialismus.

12 L.H. Sutton, "Shuffling Feet: A Discourse on the University of Göttingen," in Heinemann, "Hochschuloffiziere 1," 131–132.

13 Heinemann, "Hochschuloffiziere 2," 89.

14 James Mark in discussion in Heinemann, "Hochschuloffiziere 2," 177.

15 Geoffrey J. Giles, Students and National Socialism in Germany (Princeton: Princeton University Press, 1985).

16 Heinemann, "Hochschuloffiziere 2," 14.

17 Gimbel, "Science," 160.

18 Christoph Buchheim, Die Wiedereingliederung Westdeutschlands in die Weltwirtschaft 1945–1958 (Munich: R. Oldenbourg Verlag, 1990), 88.

19 Gimbel, "Science," 183–184.

20 Johannes Bähr, "Das Oberspreewerk — ein sowjetisches Zentrum für Röhren und Hochfrequenztechnik in Berlin (1945–1952)", in: Zeitschrift für Unternehmensgeschichte 1994/3, 145–165.

21 Ulrich Albrecht/Andreas Heinemann-Grüder/Arend Wellmann, Die Spezialisten. Deutsche Naturwissenschaftler und Techniker in der

Sowjetunion nach 1945 (Berlin: Dietz, 1992); Burghard Ciesla, "Der Spezialistentransfer in die UdSSR und seine Auswirkungen in der SBZ und DDR," in: Aus Politik und Zeitgeschichte (supplement of Das Parlament) B 49–50/1993, 24–31; Rainer Karlsch, Allein bezahlt? Die Reparationsleistungen der SBZ/DDR 1945–1953 (Berlin: Links-Verlag, 1993), 151–166.

22 Special Intelligence Memorandum No. 70 (AEG), Landesarchiv Berlin, Office of Military Government Berlin Sector (OMGBS), Dir. off. 4/137–2/7.

23 There is a third level that focuses on the mechanisms and effects of the technology transfer as concerns France and Great Britain, which shows great differences to the German-American case.

24 See Leslie E. Simon, German Research in World War II. An Analysis of the Conduct of Research (New York: John Wiley & Sons, 1947), 107, and Helmuth Trischler, Luft- und Raumfahrtforschung in Deutschland, 1990–1970 (Frankfurt: Campus, 1990).

25 Quoted after Trischler, 296.

26 See Paul Erker "New Technology and old Tradition. Industrial R&D, Innovation Policy and Corporate Culture at Siemens after World War II" (Paper presented at the 35th annual meeting of the Society for the History of Technology, Uppsala August 16–20, 1992), and Paul Erker "The Challenge of a New Technology: Transistor Research and Business Strategy at Siemens and Philips, in: History and Technology (1994) *11*: 131–43.

27 Raymond G. Stokes "Technology and the West German Wirtschaftswunder", in: Technology and Culture (1991) *32*: 1–22.

28 See ibid., 11, and Reinhard Neebe, "Technologietransfer und Außenhandel in den Anfangsjahren der Bundesrepublik Deutschland", in: Vierteljahresschrift für Sozial- und Wirtschaftsgeschichte (1989) *76*: 49–75.

29 Quoted after NARA, RG 260, OMGUS Siemens report 1946, Vol. I, 155 and 156.

30 Industrial Electrical Research in Germany, BIOS report No. 1388, 29.

31 See Siegfried Zielinski, Zur Geschichte des Videorekorders (Berlin: V. Spiess Verlag 1986), and Richard S. Rosenbloom (ed.) "Ampex Corporation and Video Innovation", in: Research on Technological Innovation, Management and Policy (1982) 2: 113–185.

32 See: Volker R. Berghahn, Paul Friedrich, Otto A. Friedrich, ein politischer Unternehmer (Frankfurt: Campus, 1993), 15ff.

33 Gimbel, Science, 183.

Index